# An Introduction to Linear Programming and Matrix Game Theory

## M. J. Fryer

Department of Mathematics,
University of Essex

**A HALSTED PRESS BOOK**

**JOHN WILEY & SONS**
**New York**

First published 1978
by Edward Arnold (Publishers) Ltd.
London.

Published in the U.S.A.
by Halsted Press, a Division
of John Wiley & Sons, Inc.,
New York.

Library of Congress Cataloging in Publication Data

Fryer, Michael John.
An introduction to linear programming and matrix
'A Halsted Press book'.
Includes index.
1. Game theory. 2. Matrices. 3. Linear programming. I Title.
QA269.F79 1977      519.7'2      77-13371
ISBN 0—470—99327—8

Printed in Great Britain

# Preface

This book has evolved from a course of lectures I originally gave to second
year undergraduate social scientists some five or six years ago.   I could
assume the students to have only a very basic (manipulative) understanding
of the algebra of simultaneous linear equations and the geometry of the line
and plane.   The course took the traditional lecture and class framework
and, as usual it was quite obvious that whereas some students avoided the
classes, others relied heavily upon them.   In preparing this book I wanted
to try to satisfy both these learning patterns.   A union of the advantages
of a 'programmed text' and the traditional course text seemed desirable and
so I decided on the present semi-programmed layout of three *distinct* parts:
Text;  Problems; Solutions.   This ensures that the book can be used as a
supervised text and is consequently also suitable for revision and reference
purposes.   The text mostly consists of fairly long *'frames'* (including
worked examples), each with an associated set of problems (designed to check
the *understanding* of the text) and a very *detailed* solution to each problem
set.   This layout seems also to give it most of the advantages of a 'linear
programmed text', and so is suitable for those students wishing to work
unsupervised at their own speed.

TO THE STUDENT

If you can solve simultaneous equations such as

$$3x + 2y = 7$$

$$-x + y = 1$$

and also represent them graphically, then you should be able to follow this
course unsupervised.

The text should be read until a directive to attempt the related problems is
encountered:  the corresponding problems section should then be worked.
Your answers to the problems should be *written out* before the solutions are
checked in the third part of the book.   If your answers are found to be
correct, continue to the next section of the text;  otherwise re-read the
current text, checking where you went wrong.

TO THE TEACHER

The first chapter introduces most of the ideas involved in obtaining a
solution to a linear programming problem by graphical methods.   The next
chapter considers the same problem but in algebraic terms, thus reinforcing
the knowledge already gained and avoiding the usual mental block associated
with purely algebraic reasoning.   Chapter 3 discusses the simplex tabular
form (with arithmetic checks), replacing the usual 'z-c' row for the
numerically equivalent, but logically simpler, updated profit row.   The
problems of starting a simplex tabular solution and of degeneracy are next
considered.   Two methods of solving the transportation problem are discussed
in Chapter 5:  the 'stepping-stone' method, which gives a real insight into

what is happening, and Dantzig's method for its computational convenience. The difficult concept of duality arises quite naturally out of the game theory section and so is postponed to the last chapter. An introduction to the terminology of matrix game theory is contained in Chapter 6, and graphical methods for solving 2 x N two person zero-sum matrix games are obtained in the following chapter. The relationships between m x n zero-sum matrix games and linear programming are discussed in the final chapter, including the all important concept of duality.

ACKNOWLEDGEMENTS

I should like to thank my colleagues Dr.R.D. Lee, Mr.T. Sprinks (both of Essex University) and Dr.F.E. Clifford (of Sussex University) for reading the manuscript, finding errors and suggesting improvements. Of course, I accept full responsibility for those errors which inevitably remain. I am also indebted to the Principal of Storrs Agricultural College, University of Connecticut, for permission to include results published in *The Competitive Position of the Connecticut Poultry Industry* by G.G. Judge.

1977                                                                M.J. Fryer

# Contents

**vi Contents**

# Chapter 1

## 1.1    INTRODUCTION

Throughout our lives we are faced with restrictions on our activities, whether personal or corporate.   One such activity which most of us face at least once in our lives is that of buying a house.   Before we can even start looking at prospective properties we have to specify our requirements and limitations:  the house must be large enough for our family yet it must cost no more than we can afford; it mustn't be too far from work or the shops or the school, etc.   Having fixed these prerequisites we can start reading the 'For Sale' advertisements.   In a sellers' market there may be no suitable houses, in which case we either give up, wait, or relax our conditions.   In a buyers' market, however, there may well be many houses satisfying our criteria and so we must decide on the best available for our purposes.   This is a typical programming problem.   Firstly there is a set of constraints.   If these cannot be satisfied we have posed an insoluble problem; if there is just one solution, the problem is solved; however, if there are several possible solutions we have a further decision problem to solve.   One way of making a rational decision in our example would be to award 'points' to each house according to its good features, and then to purchase the one with the highest number of points.   We are then optimising a preference (or utility) function with respect to constraints.   When both the constraints and function to be optimised can be expressed in mathematical terms, we have a mathematical programming problem, the type we shall be considering.

The idea of optimising a mathematical function subject to (mathematical) constraints was tackled by Lagrange over a century ago, but he was concerned solely with constraints which were in the form of exact equations.   In mathematical programming, however, the constraints are usually *inequalities*. For example, we would not look for a house costing exactly £A, but one in the price range £B to £C; this we can write as

$$B \leqslant \text{cost} \leqslant C*$$

In this book, we shall be concerned only with that subset of mathematical programming - called linear programming - in which both the expression to be optimised and all the constraints are *linear* in the variables involved, that is, if z, y, x, ... are variables (unknowns) and a, b, c, ... constants (known), then all the expressions involved must be of the form

$$a + bz + cy + dx + \ldots$$

and there must be no terms of the form

$$ax^2, \ axy, \ ax^{\frac{1}{2}}, \ \text{etc.} \ \ldots$$

This might look unduly restrictive, but in practice many everyday problems

*Read 'B *less than or equal to* cost *less than or equal to* C'

in business, engineering, management, etc. can be so formulated, even if only approximately.

Attempt problems 1.1.

## 1.2 THE PROBLEM

Rather than just quoting the rules for solving linear programming problems, we shall try to give you an insight into how the method works (and what has happened when it apparently fails). To this end we shall consider one particular problem - the theory being most easily explained in terms of a worked example - and solve it three times: graphically, then algebraically (working parallel to the graphical method), and finally in the conventional tabular method, which is little more than a shorthand form of the algebraic method. Of course, to solve a problem graphically we are limited to two (or possibly three) variables, but the other methods are not restricted in this way.

Consider the following simple two-variable problem.

A manufacturer of electronic instruments produces two types of timer: a standard and a precision model with net profits of £10 and £15 respectively. They are similar in design, each taking about the same amount of time to assemble. Let's suppose the manufacturer wishes to maximise his net profit each day subject to the availability of resources and marketing considerations. For the sake of simplicity let's consider just one such constraint: his work force can produce no more than 50 instruments per day.

It takes no expert in linear programming to realise that 50 precision timers should be produced per day to maximise the net profit when this single constraint is imposed, but how could we come to this conclusion graphically? Firstly we must formulate the problem in mathematical terms.

Supposing that a day's net profit is £P, corresponding to the manufacture of x standard and y precision instruments with profits of £10 and £15 respectively, then we have

$$P = 10x + 15y \qquad (1)$$

The constraint that *no more than* 50 instruments may be produced per day is then:

$$x + y \leqslant 50 \qquad (2)$$

There are further constraints implied in the problem which must be stated before the formalisation is complete. Assuming that a negative number of instruments will not be produced, we have

$$x \geqslant 0 \text{ and } y \geqslant 0 \qquad (3)$$

Also, both x and y should really be integers, but as this restriction is rather difficult to cope with mathematically, we shall discard it for the moment and consider the more general case in which x and y can take all values.*

We can now graph the problem using the conventional x-y rectangular

---

*Provided the values for x and y are not constrained too closely (e.g. not restricted to 0 and 1), we should be able to optimise the integer problem (approximately perhaps), by taking integer values for x and y close to the optimum obtained for the generalised problem. The numbers in this problem have been chosen so as to give integer results to avoid this difficulty.

co-ordinate system.   We start by finding all the points which satisfy the constraints, or rather, eliminating from the graph all those points which do not satisfy them.   The last two constraints (3) are the easiest to plot. Now, $x \geqslant 0$ is the set of points on and to the right of the line $x = 0$ (the y-axis), so we eliminate those points to the left of it (Fig. 1(a)). Similarly we must eliminate those points below $y = 0$ (Fig. 1(b)).   As for $x+y \leqslant 50$, this must represent all points on and to one side of the line $x+y = 50$.   If we plot this line (Fig. 1(c)), we can test which side of it is satisfied by the inequality.   The origin ($x = 0$, $y = 0$) is the easiest point to use, and this satisfies (2) (since $x+y = 0 \leqslant 50$), so we eliminate those points on the opposite side of the line to the origin.

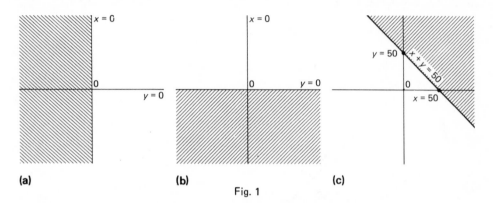

(a)                              (b)                              (c)

Fig. 1

These three constraints must be satisfied simultaneously, so we plot all the prohibited regions on the same graph (Fig. 2).   The triangular shaped 'hole' then consists of all points satisfying every constraint.

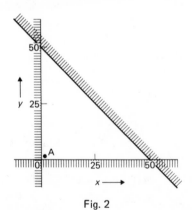

Fig. 2

This 'hole' is called the *feasible region*, and any point within it is a *feasible solution* to our problem.   For example, the point A ($x = 1$, $y = 2$) lies in the region and corresponds to a profit of $P = 10(1) + 15(2) = £40$ (from equation (1)).   Our next problem is to decide on the point (or points) which actually maximise P.

Attempt problems 1.2.

1.3   THE PROFIT LINE

Consider the equation of the profit line

## 4 The Profit Line

$$P = 10x + 15y$$

We can only plot this line once the value of P is known. Let's try various values for P and see the results. Take P = 90 initially; the line is 90 = 10x + 15y, which consists of all the points leading to a profit of £90 (see Fig. 3). However, only those of the points that are within the feasible region are possible solutions to our problem (corresponding to the solid line). If we now take P = 600, we find that this line is parallel to the first, but further away from the origin. Again, only the solid part of the line corresponds to feasible solutions. If we take P = 990 next, we have a line parallel to the others but completely outside the feasible region region. This means that there is no feasible solution to the problem resulting in a profit of £990.

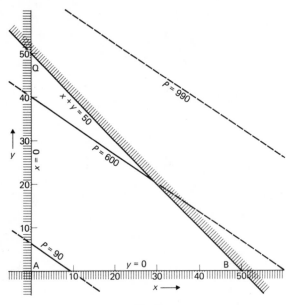

Fig. 3

Two features of this discussion are worth noting:

(i)  All the profit lines are parallel, each having the same gradient (slope) of -10/15 = -2/3.

(ii)  The further the line is from the origin, the higher the corresponding value of P.

These suggest that we shall solve the problem (i.e. find feasible values for x and y resulting in the maximum value for P) by taking that line of gradient -2/3 which is furthest from the origin and contains at least one point in the feasible region; that is, we take the line through Q with equation 10x + 15y = $P_Q$. The only feasible point on this line is Q itself, the point x = 0, y = 50, resulting in a profit $P_Q$ of £750. This agrees with the intuitive solution.

Attempt problems 1.3.

## 1.4    ADDITIONAL CONSTRAINTS

In practice, there would be many other constraints to be taken into account: for example, there might not be enough of a particular component in stock to make all precision instruments, or the sales staff might require a minimum number of standard instruments immediately to fulfil existing orders. These extra constraints might well make an intuitive solution impossible, but we can still use the graphical method.

Let's consider some extra constraints and see their effect on the profit. Suppose that there are four main components in short supply (a, b, c and d), and that they are used in different quantities for the two types of timer as shown in Table 1.    Further, suppose that on the day in question the stock inventory of these items is as in that table.

### TABLE 1  THE STOCK CONSTRAINTS

| Component | Stock | Number used per timer Standard (x) | Precision (y) |
|---|---|---|---|
| a | 220 | 4 | 2 |
| b | 160 | 2 | 4 |
| c | 370 | 2 | 10 |
| d | 300 | 5 | 6 |

The first of these restrictions is equivalent to

$$4x + 2y \leq 220$$

since each of the x standard timers uses 4 of component a, and each of the y precision ones uses 2, and the factory cannot use more than is in stock.

This constraint can be written more simply as

$$2x + y \leq 110 \tag{4}$$

Similar reasoning for the other components gives

$$2x + 4y \leq 160 \quad \text{or} \quad x + 2y \leq 80 \tag{5}$$
$$2x + 10y \leq 370 \quad \text{or} \quad x + 5y \leq 185 \tag{6}$$
$$\text{and} \quad 5x + 6y \leq 300 \tag{7}$$

Adding these constraints to the graph shows the effect on the feasible region (Fig. 4) – obviously they can't make it larger since further restrictions cannot improve one's choice.

The feasible region now consists of the area within (and those points on) the polygon ABCDE.    It is obvious that the addition of further constraints could make the feasible region vanish, in which case there would be no solution to the problem at all.

Attempt problems 1.4.

## 1.5    VARIATION OF THE PROFIT

With the profit line P = 10x + 15y and arguing as before, the maximum is found to occur at the point C.    Now suppose the profit on the standard timer were to be changed, the question would arise as to by *how much* it could be changed and the present solution (x = 20, y = 30) still remain optimum (although the optimum *value* of P might well change).    By

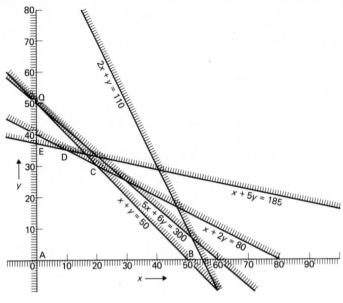

Fig. 4

considering Fig. 4 we can see that the profit line will always 'end up' at C provided its slope lies between those of the lines DC and CB. Let's suppose it becomes the same as that of DC ($-\frac{1}{2}$). The profit line is then P = 7.5x + 15y, and the optimum occurs *anywhere* along DC, so that (x = 20, y = 30) and (x = 10, y = 35) – and any point in between – all give the same maximum value for P of £600. Suppose the slope is varied again by reducing the profit on the standard model to £7, giving P = 7x + 15y (slope – 7/15); we find the optimum has now moved to the point D(x = 10, y = 35) with a corresponding maximum profit of £595.

Attempt problems 1.5.

1.6   THE VERTEX SOLUTION

One important result that emerges from this graphical presentation is that an optimum must occur at a *vertex* of the feasible region, which in two dimensions is always a *convex* polygon. ('Every chord that is not a boundary line meets the polygon in only two points'. See Fig. 5.)

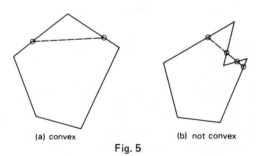

(a) convex          (b) not convex

Fig. 5

An optimal solution might occur at two neighbouring vertices, but if so, any point on the line joining them is also optimal. This means that in order to maximise P, we need only look at its values on the sides of the polygon, and in particular at its vertices. A feasible solution *at* a vertex is called *basic*. Another result worthy of note is that from any vertex of the polygon (other than that corresponding to the maximum) there is always at least one edge along which the profit does not decrease.

It is these results, intuitively obvious in two dimensions, but extending to higher dimensions, that are used in the *simplex* method for solving linear programming problems. (A simplex is just the generalisation of a convex polygon to higher dimensions. A cube is an example of a three-dimensional simplex.)

Attempt problems 1.6.

## 1.7   SUMMARY OF CHAPTER 1

A linear programming problem is one in which the expression to be optimised is linear, as are the constraints which usually take the form of inequalities.

Solutions can be obtained by graphical methods when only two (or possibly three) variables occur.

The feasible region (when it exists) is in the form of a convex polygon, and each vertex corresponds to a basic feasible solution.

An optimal solution occurs at a vertex. If optimal solutions occur at two vertices then any point on the line joining them corresponds to an optimal solution.

Attempt problems 1.7.

# Chapter 2

## 2.1 THE ALGEBRAIC SIMPLEX METHOD

This method relies on the results discussed in the first chapter and consists of the following steps:

(i)   Find any vertex of the feasible region for use as a starting point (an *initial basic feasible solution*).

(ii)  Evaluate the profit function at that point.

(iii) Choose a boundary line of the feasible region leading from that point which will result in an increased profit.

(iv)  Move to the next vertex along the chosen boundary line.

(v)   Repeat steps (ii) to (iv) until no further increase in profit can be obtained.

These steps are trivially easy to follow in our two-dimensional problem, since not only can we visualise it but also there is no decision to be made corresponding to step (iii) - once we have decided on an initial point and the direction of travel, we just keep on around the boundary until the optimum is attained.   In higher dimensions, however, there are many possible paths along the edges to the vertex at which the optimum occurs, and this method is designed not only to get there, but to follow a relatively short path.   To this end, at step (iii) we *usually* choose that boundary line giving the greatest rate of increase in the profit.   If there is a choice, we usually choose at random.   The simplex method makes use of the fact that a boundary of the feasible region corresponds to the inequality of one constraint becoming an equality (e.g. along CD (Fig.4) we have $x + 2y = 80$) whilst all the other constraints are satisfied usually as strict inequalities.   It follows that at a vertex, the intersection of (at least) two boundary lines (in two dimensions), there must be (at least) two equalities amongst the constraints (e.g. at C we have $x + 2y = 80$ and $x + y = 50$).   To keep a check on when these inequalities become equalities, we use a system of subsidiary (non-negative) *slack* variables, whose values are the differences between the left- and right-hand sides of the inequality constraints.   The constraint $x + 2y \leqslant 80$ might have the associated slack variable, t, defined by $x + 2y + t = 80$ and $t \geqslant 0$, so that the boundary CD can be denoted simply by $t = 0$, rather than the formally equivalent $x + 2y = 80$.   At points in the feasible region not on this boundary, we must have $t > 0$ (or $x + 2t < 80$).   This system of slack variables has the added advantage that the inequality constraints are replaced by equalities, the inequalities being confined to single variables being required to be non-negative.

Attempt problems 2.1

## 2.2  THE FIRST PASS

Let's rewrite the problem incorporating the slack variables r, s, t, u, v.

Maximise P = 10x + 15y  (p)

subject to

$$x + y + r = 50 \quad (a)$$
$$2x + y + s = 110 \quad (b)$$
$$x + 2y + t = 80 \quad (c)$$
$$x + 5y + u = 185 \quad (d)$$
$$5x + 6y + v = 300 \quad (e)$$
$$x, y, r, s, t, u, v \geq 0 \quad (f)$$

Algebraically our problem is now one in *seven* dimensions.  As we see from (f), all the variables are required to be non-negative.  This is natural in our problem, but it is a *general* requirement of the simplex method, and hence many problems have to have a simple modification to take this into account (as discussed at the end of the chapter).  Another general requirement is that the equations should be written so that the constant terms are not negative.

We now follow through the steps outlined in the previous section for the first time (with reference to Fig.4).

(i)   Start at a vertex of the feasible region.  The obvious one to choose is (x = 0, y = 0), the point A.  This has the advantage that the values of the slacks can be just read off equations (a) to (f):

r = 50, s = 110, t = 80, u = 185, v = 300

Note:  these values can be 'read off' simply because only one non-zero variable occurs per equation, its coefficient is unity and the constant terms are non-negative.  We shall *arrange* for this simplification to occur in the later stages of our calculations.  The equations are said to be in standard form with respect to the feasible solution (x = 0, y = 0).

(ii)  The value of P at (x = 0, y = 0) is £0.

(iii) Equation (p) shows us that each unit increase in y produces £15 extra profit, whereas a corresponding increase in x produces only £10.  *Potentially* it would seem to be most profitable for us to increase y by as much as possible.  That is, we should keep on the line x = 0, but increase y by as much as possible subject to the constraints – so we must move to the next vertex on x = 0.  (Since increasing x still increases P, we *could* have chosen to move along y = 0.)  Thus must correspond to two constraints being satisfied as equalities, x = 0 and one other (other than y = 0), but which?  Suppose we set each of them zero in turn.  This is equivalent to finding the intersections of the line x = 0 with each of the constraint boundaries

| Put | r = 0 | and then we have | y = 50 | from | (a) |
|---|---|---|---|---|---|
| | s = 0 | | y = 110 | | (b) |
| | t = 0 | | y = 40 | | (c) |
| | u = 0 | | y = 37 | | (d) |
| | v = 0 | | y = 50 | | (e) |

The values of y found are the maximum possible corresponding to their
relative constraints : for example, we cannot make y greater than 50
and still have r ⩾ 0 in (a).   So that, in order to satisfy *all* the
constraints *simultaneously*, we cannot take y greater than the least
value found.   Hence we take y = 37 and u = 0 from equation (d).   (If
two or more of the constraints had shown the value of 37 for y, it
would not then matter which we chose.)

(iv)  Our new vertex is (x = 0, u = 0), corresponding to E on the graph, as
u = 0 is the same as x + 5y = 185.

Attempt problems 2.2.

2.3    THE SECOND PASS

(ii)  Returning to step (ii) and solving x = 0, x + 5y = 185 gives
(x = 0, y = 37), so that P = £555.   Rather than substituting for x, y
in this way it will be found useful to write P in terms of the
variables which are currently to be set zero, since this will enable
us to decide which direction to take at step (iii).   Our zero
variables are currently x and u, so we substitute for y in terms of
them from (d) into (p) giving

$$P = 7x - 3u + 555 \qquad\qquad (p1)$$

confirming the profit of £555 when x = 0, u = 0.

(iii)  This new profit function shows that if we increase x by one unit we
increase P by £7, but that if we increase u by a similar amount we
actually decrease P by £3.   Obviously we must keep u = 0 (we can't
*decrease* it) but increase x by as much as possible.   If we try the
same procedure as in the first pass, setting u and one other variable
zero in turn, we get:

for y = 0;   x + r = 50, 2x + s = 110, x + t = 80, x = 185,

5x + v = 300:

for r = 0;   x + y = 50, 2x + y + s = 110, x + 2y + t = 80,

x + 5y = 185, 5x + 6y + v = 300: etc.,

giving us 5 sets of 5 simultaneous equations to solve.   The solution
of the first set is, for example, u = 0, y = 0, x = 185, r = -135,
s = -260, t = -105, v = -625, which is not a feasible solution since
we have some negative variables.   In fact, it isn't until the fourth
set of equations that we actually find a feasible point.   How much
simpler it would be if our equations (a) to (e) were in the same form
for this step as they were at the corresponding step of the first pass,
i.e.if the equations were arranged so that the non-zero variables
each appeared in one and only one equation.   If we substitute for y
in terms of x and u (the zero variables) from equation (d) this will
be found to happen, and we get

$$4x/5 \quad + r \quad - \quad u/5 \quad = 13 \qquad\qquad (a1)$$
$$9x/5 \quad + s \quad - \quad u/5 \quad = 73 \qquad\qquad (b1)$$
$$3x/5 \quad + t - 2u/5 \quad = 6 \qquad\qquad (c1)$$
$$x/5 + y \quad + \quad u/5 \quad = 37 \qquad\qquad (d1)$$
$$19x/5 \quad - 6u/5 + v = 78 \qquad\qquad (e1)$$

Equation (dl) is just equation (d) divided by 5 to make the coefficient of y unity (enabling us to read off the values of all the non-zero variables). Our equations are now again in standard form but with respect to (x = 0, u = 0).

We can now start step (iv) again, setting u = 0 and observing the maximum increase in x allowed by each equation.

| Put | r = 0 | and then we get | x = 65/4 | from | (al) |
|---|---|---|---|---|---|
| | s = 0 | | x = 365/0 | | (bl) |
| | t = 0 | | x = 10 | | (cl) |
| | y = 0 | | x = 185 | | (dl) |
| | v = 0 | | x = 390/19 | | (el) |

As before we must pick the smallest of these so as not to violate any of the constraints. So we must move to x = 10 corresponding to u = 0 and t = 0.

Attempt problems 2.3.

## 2.4 THE THIRD PASS

(ii) We are now at the vertex D where the profit is £625.

(iii) Before we can assess which variable should be increased, we must write the profit function in terms of u and t (the two variables set zero); that is, eliminate x by substituting from (cl) into (pl), giving

$$P = 625 + 5u/3 - 35t/3 \tag{p2}$$

We can now increase the profit only by increasing u.

(iv) We must also make this substitution into each of (al) to (el) to make them easily manageable.

$$r \quad - \quad 4t/3 + u/3 \quad = \quad 5 \tag{a2}$$
$$s - \quad 3t \quad + \quad u \quad = 55 \tag{b2}$$
$$x \qquad + \quad 5t/3 - 2u/3 \quad = 10 \tag{c2}$$
$$y \qquad - \quad t/3 + u/3 \quad = 35 \tag{d2}$$
$$- \quad 19t/3 + 4u/3 + v = 40 \tag{e2}$$

In order to find the maximum increase in u we set t and each of the other variables to zero, one at a time

| Put | r = 0 | and then we get | u = 15 | from | (a2) |
|---|---|---|---|---|---|
| | s = 0 | | u = 55 | | (b2) |
| | x = 0 | | u = -15 | | (c2) |
| | y = 0 | | u = 105 | | (d2) |
| | v = 0 | | u = 30 | | (e2) |

The equation (c2) has given us a *negative* result; all this really means is that it places no upper limit at all on the possible values for u. For example, if we take the smallest *positive* value allowable for u, (15 from (a2), corresponding to r = 0, t = 0) and substitute into (c2), we find

x - 10 + 0 = 10 i.e. x = 20,

which does not violate any constraint. This means we can disregard any negative values given for u at this step and concentrate on the smallest non-negative value.

Attempt problems 2.4.

2.5  THE FOURTH PASS

(ii)  The profit is now £650, but can it be increased still further?

(iii) Substituting for u in terms of r and t from (a2) gives

$$P = 650 - 5r - 5t \tag{p3}$$

This shows that no increase in P can be obtained by increasing r or t. Hence we have arrived at a solution to the problem:

P = £650, r = 0, t = 0 , x = 20, y = 30, s = 40, u = 15, v = 20.

This corresponds to the solution found graphically, but this method, although more cumbersome in two dimensions, can be used quite readily in problems whose (initial) dimensions are much higher.

In order to deal with a variable not restricted to be non-negative we use the simple strategy of replacing it by the difference of two new (non-negative) variables. For example, if x is unrestricted, we define two new variables $x_1$, $x_2 (\geqslant 0)$, and set $x = x_1 - x_2$. These variables now allow x to take any value, positive or negative.

Attempt problems 2.5.

2.6  SUMMARY OF CHAPTER 2

The simplex method requires that the constraint inequalities involving several variables be turned into equalities by the use of slack variables and that all the variables and constant terms be non-negative. It consists of the following steps:

(i)  Find an initial basic feasible solution.

(ii)  Write the profit line in terms of the variables set to zero.

(iii) Choose a variable in the profit line which has a positive coefficient. If there is a choice, it is usual to take the one with the largest coefficient. This new variable defines the boundary along which to move next.

(iv)  Write the constraints in standard form, so that the non-zero variables each appear in one and only one equation and have unit coefficients. This will indicate by how much the new variable may be increased without violating any constraints. (N.B. Disregard negative results here.)

(v)  Repeat steps (ii) to (iv) until there are no positive coefficients available at step (iii). A feasible solution optimising the value of the profit has then been found.

Attempt problems 2.6.

# Chapter 3

## 3.1    THE TABLEAU

We can now proceed to the tabular form of the simplex method.    There is
only a minor variation from the method used above, in that we write
$P - 10x - 15y = 0$ rather than $P = 10x + 15y$, since the right-hand side will
then represent the profit when the variables (excluding P) on the left-hand
side are set to zero.    The entries in Table 2 correspond to our initial
equations;    the rows and columns have been numbered for reference purposes
only.    Our initial vertex will be ($x = 0$, $y = 0$) as before.

### TABLE 2   THE INITIAL TABLEAU

|   |   | 1 | 2 | 3 | 4 | 5 | 6 | 7 | 8 | 9 | 10 |
|---|---|---|---|---|---|---|---|---|---|---|----|
| 1 |   | x | y | r | s | t | u | v |   |   |    |
| 2 | P | -10 | -15 | 0 | 0 | 0 | 0 | 0 | 0 |   |    |
| 3 | r | 1 | 1 | 1 | 0 | 0 | 0 | 0 | 50 | 50 |   |
| 4 | s | 2 | 1 | 0 | 1 | 0 | 0 | 0 | 110 | 110 |   |
| 5 | t | 1 | 2 | 0 | 0 | 1 | 0 | 0 | 80 | 40 |   |
| 6 | u | 1 | 5 | 0 | 0 | 0 | 1 | 0 | 185 | 37 |   |
| 7 | v | 5 | 6 | 0 | 0 | 0 | 0 | 1 | 300 | 50 |   |

Row 1 lists the variables, and row 2 the corresponding coefficients in the
P-equation : similarly rows 3 to 7 (cols 2-9) display the coefficients in
the constraint equations, column 9 being the *constant* column.    Column 1
(rows 3 to 7) lists the variables not set to zero initially (the *basis*).
Column 10 has yet to be calculated.    Note the use of vertical lines to
split the table into its logical parts.

Having taken ($x = 0$, $y = 0$) as our initial vertex (or *initial basic feasible
solution*), we must now decide which of these to vary and then which other
variable to set zero.    Following our previous reasoning, we look for the
coefficient in the P-row with the largest *negative* value - 15 in the y
column (col 3) - which is shown circled.    (Negative now, as the P-equation
has been re-written.)

This column is now referred to as the *pivot column*.    How many y's can we
introduce?    As before, we divide the entries in column 9 (rows 3 to 7 only)
by the corresponding entries in the pivot column and write the results in
column 10 (the *division* column).    The smallest (positive) value (37, shown
circled) corresponds to u being zero and y being 37.    This row is known as
the *pivot row*.    The element at the intersection of the pivot row and column
is known as the *pivot* (shown boxed).    The next stage is to eliminate the y
coefficient from each of the rows (equations) except the one containing the
pivot.    We do this by dividing the elements of the pivot row (equation) by

5 to reduce the y coefficient (pivot) to 1 (row 12), and then subtracting
multiples of this row from the other rows to eliminate the y's.   As new
rows are formed, we set them out in the second tableau (Table 3).   For
example, row 9 (replacing row 3) is the original row 3 minus row 12, and row
13 (replacing row 7) is the original row 7 minus 6 times row 12, etc.   We
omit the list of variables.

### TABLE 3   THE SECOND TABLEAU

|   |   | 1 | 2 | 3 | 4 | 5 | 6 | 7 | 8 | 9 | 10 |
|---|---|---|---|---|---|---|---|---|---|---|----|
| 8 | P | $\boxed{-7}$ | 0 | 0 | 0 | 0 | 0 | 3 | 0 | 555 |   |
| 9 | r | $4/5$ | 0 | 1 | 0 | 0 | $-1/5$ | 0 | 13 | $65/4$ |
| 10 | s | $9/5$ | 0 | 0 | 1 | 0 | $-1/5$ | 0 | 73 | $365/9$ |
| 11 | t | $\boxed{3/5}$ | 0 | 0 | 0 | 1 | $-2/5$ | 0 | 6 | $\boxed{10}$ |
| 12 | y | $1/5$ | 1 | 0 | 0 | 0 | $1/5$ | 0 | 37 | 185 |
| 13 | v | $19/5$ | 0 | 0 | 0 | 0 | $-6/5$ | 1 | 78 | $390/19$ |

The new non-zero variable, y, must now replace u in column 1 (row 12).   We
say that y has replaced u in the basis.   The new P-row (row 8) is formed
just as the other rows, showing that the profit is now £555 and indicating
that x should be increased.   Column 2 is therefore the new pivot column.
Column 10, obtained by dividing the elements of column 9 by corresponding
ones in the pivot column, shows that the new pivot row is 11.

A check with equations (a1) to (e1) and the results of stage (iv) of the
second pass will show that this second tableau contains all the information
previously obtained but in a much more compact form.  We continue adding
tableaux until no elements in the P-row are negative.   The complete set of
tableaux is shown in Table 4.

This last tableau shows us that an optimal solution is given by:

x = 20, y = 30, r = 0, s = 40, t = 0, u = 15, v = 20

corresponding to a profit of £650.   This is the same answer as we obtained
before, and moreover, from our graphical solution we know it to be unique.

Attempt problems 3.1

### 3.2   CHECKING PROCEDURE

By now it must be all too obvious that arithmetical accuracy is all-
important when solving linear programming problems by hand.   A single slip
in an early tableau could make the 'solution' non-optimal and yet not always
nonsensical enough to be discarded.   There is a relatively simple system of
checks that can be built into the tabular method, which should eliminate
most arithmetic slips without increasing the working overmuch.   It requires
that a further column be added to the right of the table, the check column,
and in it placed the sum of the elements in each row of the tableau,
including the 'constant' column but excluding the 'division' column.   So
that, for example, the first tableau of the previous section would be as in
Table 5.

## TABLE 4  A COMPLETE SOLUTION

|   | x | y | r | s | t | u | v |   |   |
|---|---|---|---|---|---|---|---|---|---|
| P | -10 | (-15) | 0 | 0 | 0 | 0 | 0 | 0 |   |
| r | 1 | 1 | 1 | 0 | 0 | 0 | 0 | 50 | 50 |
| s | 2 | 1 | 0 | 1 | 0 | 0 | 0 | 110 | 110 |
| t | 1 | 2 | 0 | 0 | 1 | 0 | 0 | 80 | 40 |
| u | 1 | [5] | 0 | 0 | 0 | 1 | 0 | 185 | (37) |
| v | 5 | 6 | 0 | 0 | 0 | 0 | 1 | 300 | 50 |
| P | (-7) | 0 | 0 | 0 | 0 | 3 | 0 | 555 |   |
| r | $4/5$ | 0 | 1 | 0 | 0 | $-1/5$ | 0 | 13 | $65/4$ |
| s | $9/5$ | 0 | 0 | 1 | 0 | $-1/5$ | 0 | 73 | $365/9$ |
| t | $[3/5]$ | 0 | 0 | 0 | 1 | $-2/5$ | 0 | 6 | (10) |
| y | $1/5$ | 1 | 0 | 0 | 0 | $1/5$ | 0 | 37 | 185 |
| v | $19/5$ | 0 | 0 | 0 | 0 | $-6/5$ | 1 | 78 | $390/19$ |
| P | 0 | 0 | 0 | 0 | $35/3$ | $(-5/3)$ | 0 | 625 |   |
| r | 0 | 0 | 1 | 0 | $-4/3$ | $[1/3]$ | 0 | 5 | (15) |
| s | 0 | 0 | 0 | 1 | $-3$ | 1 | 0 | 55 | 55 |
| x | 1 | 0 | 0 | 0 | $5/3$ | $-2/3$ | 0 | 10 | -15 |
| y | 0 | 1 | 0 | 0 | $-1/3$ | $1/3$ | 0 | 35 | 105 |
| v | 0 | 0 | 0 | 0 | $-19/3$ | $4/3$ | 1 | 40 | 30 |
| P | 0 | 0 | 5 | 0 | 5 | 0 | 0 | 650 |   |
| u | 0 | 0 | 3 | 0 | -4 | 1 | 0 | 15 |   |
| s | 0 | 0 | -3 | 1 | 1 | 0 | 0 | 40 |   |
| x | 1 | 0 | 2 | 0 | -1 | 0 | 0 | 20 |   |
| y | 0 | 1 | -1 | 0 | 1 | 0 | 0 | 30 |   |
| v | 0 | 0 | -4 | 0 | -1 | 0 | 1 | 20 |   |

## TABLE 5  THE CHECK COLUMN

|   | x | y | r | s | t | u | v |   | chk |
|---|---|---|---|---|---|---|---|---|---|
| P | -10 | -15 | 0 | 0 | 0 | 0 | 0 | 0 | -25 |
| r | 1 | 1 | 1 | 0 | 0 | 0 | 0 | 50 | 53 |
| s | 2 | 1 | 0 | 1 | 0 | 0 | 0 | 110 | 114 |
| t | 1 | 2 | 0 | 0 | 1 | 0 | 0 | 80 | 84 |
| u | 1 | 5 | 0 | 0 | 0 | 1 | 0 | 185 | 192 |
| v | 5 | 6 | 0 | 0 | 0 | 0 | 1 | 300 | 312 |

## 16 Summary

The checks are calculated *before* the 'division' column is filled so that
there should be no chance of including them in the sum by mistake.    The
elements in the check column are then treated in exactly the same way as all
the other elements in the same row when calculating the second tableau.
They should then still be the sums of the elements in the rows of the new
tableau.    For example the y-row of the next tableau (Table 6) will be 1/5
of the u-row in this tableau (Table 5), so that the y-check will be 192/5,
confirming that all the entries in the y-row are correct since they do add
up to 192/5.    Similarly, the new profit row is obtained by adding to the
original profit row 15 times the y-row:   hence the P-check is
$-25 + 15 (192/5) = 551$, which is the sum of the entries in the P-row of the
second tableau.

### TABLE 6   THE CHECKING PROCEDURE

| | | | | | | | | | |
|---|---|---|---|---|---|---|---|---|---|
| P | $-7$ | O | O | O | O | 3 | O | 555 | 551 |
| r | $4/5$ | O | 1 | O | O | $-1/5$ | O | 13 | $14\,3/5$ |
| s | $9/5$ | O | O | 1 | O | $-1/5$ | O | 73 | $75\,3/5$ |
| t | $3/5$ | O | O | O | 1 | $-2/5$ | O | 6 | $7\,1/5$ |
| y | $1/5$ | 1 | O | O | O | $1/5$ | O | 37 | $38\,2/5$ |
| v | $19/5$ | O | O | O | O | $-6/5$ | 1 | 78 | $81\,3/5$ |

Attempt problems 3.2

### 3.3    SUMMARY OF CHAPTER 3

After the equations have been written in standard form corresponding to an
initial basic feasible solution, their coefficients are placed in the initial
tableau.    The largest *negative* entry in the profit line defines the pivot
column.    The pivot row corresponds to the smallest *positive* result obtained
from dividing the 'constant' column (penultimate column) by the corresponding
entries in the pivot column.    The next tableau is now started.    The pivot
element is reduced to unity and the other entries in the pivot column are
eliminated by subtracting suitable multiples of the pivot row.    The equations
are again in standard form with the pivot column variable replacing that of
the pivot row in the basis.

These steps are repeated until there are no longer any negative coefficients
in the profit line.    An optimal solution has then been found.

Attempt problems 3.3.

# Chapter 4

## 4.1 THE INITIAL SOLUTION

Suppose that in the timer production problem we introduce the further
constraint that the sales staff require a minimum of 5 standard instruments
to fulfil urgent orders. We write this as $x \geq 5$, which, if we introduce
a slack variable, $w$, becomes $x - w = 5$, $w \geq 0$. *Note the minus sign.* This
means that our previous initial basic feasible solution ($x = 0$, $y = 0$) is no
longer valid since it requires $w = -5$. One way of bypassing this difficulty
is to use $x = 5$, $y = 0$ (or equivalently $w = 0$, $y = 0$) as the initial basic
feasible solution. The elements of the tableau, however, are not then in
the standard form for the simplex method, since $x$, a non-zero variable,
occurs in all the rows (equations). We must eliminate $x$ from all the rows
except the last : this is equivalent to taking $x$ for the pivot row and
column (without referring to the P-equation). See Table 7. Then in the
second tableau we have the initial basic feasible solution ($y = 0$, $w = 0$)
and the equations in the standard form.

### TABLE 7  STANDARDISATION OF THE INITIAL SOLUTION

|   | x | y | r | s | t | u | v | w |   | chk |
|---|---|---|---|---|---|---|---|---|---|-----|
| P | -10 | -15 | 0 | 0 | 0 | 0 | 0 | 0 | 0 | -25 |
| r | 1 | 1 | 1 | 0 | 0 | 0 | 0 | 0 | 50 | 53 |
| s | 2 | 1 | 0 | 1 | 0 | 0 | 0 | 0 | 110 | 114 |
| t | 1 | 2 | 0 | 0 | 1 | 0 | 0 | 0 | 80 | 84 |
| u | 1 | 5 | 0 | 0 | 0 | 1 | 0 | 0 | 185 | 192 |
| v | 5 | 6 | 0 | 0 | 0 | 0 | 1 | 0 | 300 | 312 |
| x | 1 | 0 | 0 | 0 | 0 | 0 | 0 | -1 | 5 | 5 |
| P | 0 | -15 | 0 | 0 | 0 | 0 | 0 | -10 | 50 | 25 |
| r | 0 | 1 | 1 | 0 | 0 | 0 | 0 | 1 | 45 | 48 |
| s | 0 | 1 | 0 | 1 | 0 | 0 | 0 | 2 | 100 | 104 |
| t | 0 | 2 | 0 | 0 | 1 | 0 | 0 | 1 | 75 | 79 |
| u | 0 | 5 | 0 | 0 | 0 | 1 | 0 | 1 | 180 | 187 |
| v | 0 | 6 | 0 | 0 | 0 | 0 | 1 | 5 | 275 | 287 |
| x | 1 | 0 | 0 | 0 | 0 | 0 | 0 | -1 | 5 | 5 |

Attempt problems 4.1

## 4.2 DUMMY VARIABLES

We now introduce a general method of starting the simplex method.

## 18 Dummy Variables

Let's consider the following problem:

*Minimise*    4x + 3y

subject to    2x + y ≥ 50

x + 2y ≥ 40

5x + 4y ≥ 170

x ≥ 0, y ≥ 0

We can convert this into a maximisation problem by considering

P = -4x - 3y

and trying to *maximise* P in the usual way.

We must now write the problem in a standard form using non-negative slacks u, v, w:

Maximise        P = -4x - 3y

subject to        2x + y  - u            = 50

x + 2y    - v            = 40

5x + 4y          - w    = 170

x, y, u, v, w ≥ 0

We now have the problem of finding an initial basic feasible solution, since the coefficients of the 'natural' basis u, v, w are negative.   If no initial basic solution is at hand, the method usually employed is to introduce yet another set of non-negative *dummy* (or artificial) variables r, s, t, say, one for each equation and having the required unit coefficients. The constraints now become

2x + y - u              + r        =  50

x + 2y    - v          + s      =  40

5x + 4y        - w        + t    = 170

x, y, u, v, w, r, s, t ≥ 0

and we have the obvious initial solution x = y = u = v = w = 0, r = 50, s = 40, t = 170.    However, if we substitute this into the profit equation, we see that we have the maximum value possible, zero!    The fallacy here, of course, is connected with introducing dummy variables - they really have nothing to do with the problem, so they must be eliminated before we can arrive at a sensible solution.    The way to do this after using them to 'prime' our problem is to introduce them into the profit function with an enormous penalty, M, to ensure they are made zero (if possible) in any optimal solution.    We then have

P = -4x - 3y - M(r + s + t)

where M is large enough to dominate any expression in which it occurs. Unfortunately the P-equation is still not in the correct form, since it contains the (dummy) variables r,s,t which are not initially zero.    Our first step must therefore be to eliminate these by subtracting M times each of the constraints from the new profit function, giving

P + (4-8M)x+(3-7M)y +Mu + Mv + Mw = -260M.

The solution follows in the usual manner, as shown.    (The checks have been omitted.)

### TABLE 8  USE OF DUMMY VARIABLES

|   | x | y | u | v | w | r | s | t | | |
|---|---|---|---|---|---|---|---|---|---|---|
| P | (4-8M) | 3-7M | M | M | M | 0 | 0 | 0 | -260M | |
| r | [2] | 1 | -1 | 0 | 0 | 1 | 0 | 0 | 50 | (25) |
| s | 1 | 2 | 0 | -1 | 0 | 0 | 1 | 0 | 40 | 40 |
| t | 5 | 4 | 0 | 0 | -1 | 0 | 0 | 1 | 170 | 34 |
| P | 0 | (1-3M) | 2-3M | M | M | 4M-2 | 0 | 0 | -100-60M | |
| x | 1 | ½ | -½ | 0 | 0 | ½ | 0 | 0 | 25 | 50 |
| s | 0 | [3/2] | ½ | -1 | 0 | -½ | 1 | 0 | 15 | (10) |
| t | 0 | 3/2 | 5/2 | 0 | -1 | -5/2 | 0 | 1 | 45 | 30 |
| P | 0 | 0 | (5/3 - 2M) | 2/3 - M | M | 3M - 5/3 | 2/3 + 2M | 0 | -110-30M | |
| x | 1 | 0 | -2/3 | 1/3 | 0 | 2/3 | -1/3 | 0 | 20 | NEG |
| y | 0 | 1 | 1/3 | -2/3 | 0 | -1/3 | 2/3 | 0 | 10 | 30 |
| t | 0 | 0 | [2] | 1 | -1 | -2 | -1 | 1 | 30 | (15) |
| P | 0 | 0 | 0 | (-1/6) | 5/6 | M | 1/6 + M | M - 5/6 | -135 | |
| x | 1 | 0 | 0 | 2/3 | -1/3 | 0 | -2/3 | 1/3 | 30 | 45 |
| y | 0 | 1 | 0 | -5/6 | 1/6 | 0 | 5/6 | -1/6 | 5 | NEG |
| u | 0 | 0 | 1 | [½] | -½ | -1 | -½ | ½ | 15 | (30) |
| P | 0 | 0 | 1/3 | 0 | 2/3 | -1/3 + M | M | M - 2/3 | -130 | |
| x | 1 | 0 | -4/3 | 0 | 1/3 | 4/3 | 0 | -1/3 | 10 | |
| y | 0 | 1 | 5/3 | 0 | -2/3 | -5/3 | 0 | 2/3 | 30 | |
| v | 0 | 0 | 2 | 1 | -1 | -2 | -1 | 1 | 30 | |

The solution is P = -130 (Cost = 130) x = 10, y = 30, v = 30, u = 0, w = 0 and r = s = t = 0 as required.

There are rather more calculations here than are really necessary.    We know that r,s,t must not appear in the final solution, so that once any one of them has been removed from the basis, the magnitude of M will preclude it from ever returning.    This means that we need no longer calculate the co-efficients of any dummy variables once they have left the basis.    The unnecessary working is shown in the dotted region of the tableau.    (N.B. Remember to adjust the check column accordingly!)

Attempt problems 4.2.

4.3.   DEGENERACY

Let's consider the original timer-production problem with the alternative extra constraint

$7x + 20y \leqslant 740$, or equivalently

$7x + 20y + w = 740$, $w \geqslant 0$

if we add it to the graph, (Fig. 6) we see that its boundary cuts a corner off the original feasible region, since it passes through the two existing vertices C and E.

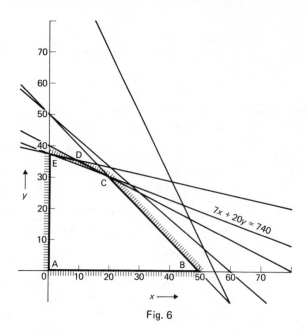

Fig. 6

At each of these two (*degenerate*) vertices three boundary lines intersect, rather than the usual two needed to define the vertex in two dimensions. This complicates matters slightly, since the simplex method relies on there being only two zero variables at any one vertex (in two dimensions, that is). On coming to such a vertex it can't decide which of the two other boundaries should be chosen next. Fortunately, if it chooses incorrectly the fault is usually rectified at the next tableau.* For example, in travelling through E, we have $x = 0$, $y = 0$ initially and then the choice of taking $z = 0$ and $u = 0$ or $x = 0$ and $w = 0$. If we decide on the former, the next tableau forces us to take $u = 0$ and $w = 0$, so that in either case we leave E along $w = 0$. The working for the longer of these alternatives is set out in Table 9.

Note that in deciding on the pivot row we must distinguish between $+0$ and $-0$ (dividing zero by positive and negative numbers respectively), the latter being disregarded. Only the coefficients in the constant column of a last tableau *need* be calculated, although a complete evaluation serves as a useful check.

*In some circumstances it is possible that the solution will 'cycle' around several degenerate solutions and never reach the optimum. Fortunately *practical* problems never seem to exhibit this feature, although several artificial problems have been published which do so. A simple way of ensuring that cycling will not prevent the optimum ever being attained is to choose the pivot row at random whenever there is a choice.

### TABLE 9   THE DEGENERATE SOLUTION

| | | x | y | r | s | t | u | v | w | | | chk |
|---|---|---|---|---|---|---|---|---|---|---|---|---|
| **A** | P | -10 | (-15) | 0 | 0 | 0 | 0 | 0 | 0 | 0 | | -25 |
| | r | 1 | 1 | 1 | 0 | 0 | 0 | 0 | 0 | 50 | 50 | 53 |
| | s | 2 | 1 | 0 | 1 | 0 | 0 | 0 | 0 | 110 | 110 | 114 |
| | t | 1 | 2 | 0 | 0 | 1 | 0 | 0 | 0 | 80 | 40 | 84 |
| | u | 1 | [5] | 0 | 0 | 0 | 1 | 0 | 0 | 185 | (37) | 192 |
| | v | 5 | 6 | 0 | 0 | 0 | 0 | 1 | 0 | 300 | 50 | 312 |
| | w | 7 | 20 | 0 | 0 | 0 | 0 | 0 | 1 | 740 | 37 | 768 |
| **E** | P | (-7) | 0 | 0 | 0 | 0 | 3 | 0 | 0 | 555 | | 551 |
| | r | $\frac{4}{5}$ | 0 | 1 | 0 | 0 | $-\frac{1}{5}$ | 0 | 0 | 13 | $\frac{65}{4}$ | $\frac{73}{5}$ |
| | s | $\frac{9}{5}$ | 0 | 0 | 1 | 0 | $-\frac{1}{5}$ | 0 | 0 | 73 | $\frac{365}{9}$ | $\frac{378}{5}$ |
| | t | $\frac{3}{5}$ | 0 | 0 | 0 | 1 | $-\frac{2}{5}$ | 0 | 0 | 6 | 10 | $\frac{36}{5}$ |
| | y | $\frac{1}{5}$ | 1 | 0 | 0 | 0 | $\frac{1}{5}$ | 0 | 0 | 37 | 185 | $\frac{192}{5}$ |
| | v | $\frac{19}{5}$ | 0 | 0 | 0 | 0 | $-\frac{6}{5}$ | 1 | 0 | 78 | $\frac{390}{19}$ | $\frac{408}{5}$ |
| | w | [3] | 0 | 0 | 0 | 0 | -4 | 0 | 1 | 0 | (+0) | 0 |
| **E** | P | 0 | 0 | 0 | 0 | 0 | $\left(-\frac{19}{3}\right)$ | 0 | $\frac{7}{3}$ | 555 | | 551 |
| | r | 0 | 0 | 1 | 0 | 0 | $\frac{13}{15}$ | 0 | $-\frac{4}{15}$ | 13 | (15) | $\frac{73}{5}$ |
| | s | 0 | 0 | 0 | 1 | 0 | $\frac{11}{5}$ | 0 | $-\frac{3}{5}$ | 73 | $\frac{365}{11}$ | $\frac{378}{5}$ |
| | t | 0 | 0 | 0 | 0 | 1 | $\frac{2}{5}$ | 0 | $-\frac{1}{5}$ | 6 | 15 | $\frac{36}{5}$ |
| | y | 0 | 1 | 0 | 0 | 0 | $\frac{7}{15}$ | 0 | $-\frac{1}{15}$ | 37 | $\frac{555}{7}$ | $\frac{192}{5}$ |
| | v | 0 | 0 | 0 | 0 | 0 | $\frac{58}{15}$ | 1 | $-\frac{19}{15}$ | 78 | $\frac{585}{29}$ | $\frac{408}{5}$ |
| | x | 1 | 0 | 0 | 0 | 0 | $-\frac{4}{3}$ | 0 | $\frac{1}{3}$ | 0 | -0 | 0 |
| **C** | P | 0 | 0 | $\frac{95}{13}$ | 0 | 0 | 0 | 0 | $\frac{5}{13}$ | 650 | | $657\frac{9}{13}$ |
| | u | 0 | 0 | $\frac{15}{13}$ | 0 | 0 | 1 | 0 | $-\frac{4}{13}$ | 15 | | $\frac{219}{13}$ |
| | s | 0 | 0 | $-\frac{33}{13}$ | 1 | 0 | 0 | 0 | $\frac{1}{13}$ | 40 | | $\frac{501}{13}$ |
| | t | 0 | 0 | $-\frac{6}{13}$ | 0 | 1 | 0 | 0 | $-\frac{1}{13}$ | 0 | | $\frac{6}{13}$ |
| | y | 0 | 1 | $-\frac{7}{13}$ | 0 | 0 | 0 | 0 | $\frac{1}{13}$ | 30 | | $\frac{397}{13}$ |
| | v | 0 | 0 | $-\frac{58}{13}$ | 0 | 0 | 0 | 1 | $-\frac{1}{13}$ | 20 | | $\frac{214}{13}$ |
| | x | 1 | 0 | $\frac{20}{13}$ | 0 | 0 | 0 | 0 | $-\frac{1}{13}$ | 20 | | $\frac{292}{13}$ |

Annotations:

- (Block A, t row) Choice of pivot row
- (Block A, u row) ← We arbitrarily choose to leave along u = 0.
- (Block A, w row) ←
- (Block E, first) Degenerate solution since one of basis variables also zero.
- (Block E, w row) ← zero.
- (Block E, second, P row) ← Profit unchanged.
- (Block E, second, r row) ←
- (Block E, second, s row) Choice
- (Block E, second, t row) ←
- (Block E, second, x row) ← solution   Degenerate
- (Block C, P row) ← Optimum
- (Block C, s row) Degenerate
- (Block C, t row) ← solution

Attempt problems 4.3.

## 4.4 THE OPTIMAL SOLUTION

As can be seen from the graphical approach, the simplex technique aims at finding an optimal solution to a linear programming problem (if one exists) by following a relatively short path along boundary lines. If, however, there are several optimal solutions, the simplex method finds one and indicates that there may be others by showing a zero element in the P-row which does *not* correspond to a variable already in the basis. The implication is that this other variable may be brought into the basis without changing the profit.

On the other hand, when the feasible region contains no points - i.e. no solutions exist - the simplex technique clearly indicates the fact by not allowing the dummy variables to be dropped from the final solution.

The other non-standard form of solution that occurs is when one of the variables becomes infinite, as might be the case when the feasible region is 'open ended'. This type of result is indicated by a negative element in the P-row, and every finite element in the division column being negative (implying that at least one variable can be increased without limit).

Attempt problems 4.4.

## 4.5. SUMMARY OF CHAPTER 4

When the constraints are in the form "≤" (and the right-hand sides are non-negative), the initial basic feasible solution is usually obtained by taking the slacks as the basis. When the constraints are "≥" or "=", the addition of dummy variables allows us to take them as the initial basis, provided they are excluded from the final (optimal) solution by their inclusion in the profit line with a large penalty. Degeneracy occurs when more than the number of constraint boundaries required to define a vertex meet at that vertex.

Attempt problems 4.5

# Chapter 5

## 5.1 FURTHER ALGORITHMS

You should by now have a reasonable understanding of the standard simplex technique and the type of problem which it can handle. There is, however, an alternative tabular form, the revised simplex technique, which handles the same type of problem, but in an even more compact way. For this reason it is less suited to hand calculation and more to manipulation by computers, and we shall not discuss it further.

The standard simplex technique is therefore really restricted in its use to problems which can be solved by hand (less than 10 variables or constraints, say). However, there is an adaptation of the technique which works well (by hand) even with large numbers of variables and constraints (up to 100?) in a certain class of problems, known as transportation problems. One such problem might be to find the cheapest way of distributing goods to retailers from wholesale depots when limited stocks are available.

*The Transportation Problem*

As with the derivation of the standard simplex technique, we shall consider the following typical transport problem with which to develop the method. A company which manufactures electricty generators has three distribution depots situated in the Midlands. When a strike of power workers is announced one winter, large orders are placed by four Midlands' retailers, which would use up between them all the available stocks. The manufacturer needs to know how the orders should be met if he is to minimise his transport costs. Suppose all the information is as shown in Table 10, where the costs given are the transportation costs of sending one generator from a given depot to a given retailer.

### TABLE 10  TRANSPORTATION DATA

|  | | Depots | | |
|---|---|---|---|---|
| Unit Costs | $D_1$ | $D_2$ | $D_3$ | Orders |
| Retailers $R_1$ | 17 | 16 | 14 | 31 |
| $R_2$ | 11 | 14 | 13 | 32 |
| $R_3$ | 15 | 11 | 14 | 45 |
| $R_4$ | 12 | 11 | 10 | 20 |
| Stocks | 43 | 55 | 30 | 128 |

Before we can proceed we must show that this is indeed a linear programming problem. Suppose

$D_1$ sends $x_1$ generators to $R_1$, $x_2$ to $R_2$, etc.

$D_2$ sends $y_1$ generators to $R_1$, $Y_2$ to $R_2$, etc.

$D_3$ sends $z_1$ generators to $R_1$, $z_2$ to R , etc.

then the transportation costs can be written as

$$C = 17x_1 + 16y_1 + 14z_1 + 11x_2 + 14y_2 + 13z_2 + 15x_3 + 11y_3 +$$

$$14z_3 + 12x_4 + 11y_4 + 10z_4 \qquad\qquad\text{(c)}$$

subject to the 'stock' constraints

$$x_1 + x_2 + x_3 + x_4 \qquad\qquad = 43 \qquad\text{(1)}$$

$$y_1 + y_2 + y_3 + y_4 \qquad\qquad = 55 \qquad\text{(2)}$$

$$z_1 + z_2 + z_3 + z_4 = 30 \qquad\text{(3)}$$

the 'order' constraints

$$x_1 \qquad\qquad + y_1 \qquad\qquad + z_1 \qquad\qquad = 31 \qquad\text{(4)}$$

$$x_2 \qquad\qquad + y_2 \qquad\qquad + z_2 \qquad = 32 \qquad\text{(5)}$$

$$x_3 \qquad\qquad + y_3 \qquad\qquad + z_3 \quad = 45 \qquad\text{(6)}$$

$$x_4 \qquad\qquad + y_4 \qquad\qquad + z_4 = 20 \qquad\text{(7)}$$

and the constraint that no variable may be negative.

Here we are trying to optimise a linear function of twelve variables subject to seven constraints - a linear programming problem. Note, however, that the constraint equations are not independent, since the sum of the first three is the same as the sum of the last four: there are only six *independent* constraints, so that for a basic feasible solution we must have just six non-zero (basis) variables. The problem is *not* in standard form, but in this modified technique there is no need for it to be: we must just remember which variables are in the basis at any one time. The mathematical simplicity of the constraints in the transportation problem is self-evident: all the coefficients are either 0 or 1, and there is obvious symmetry. This simplicity is reflected in their solution: there is always a basic feasible solution to the problem and such a solution can always be found *by inspection*! Hence we don't have to worry about dummy variables for priming the problem. Also there are several relatively efficient (i.e. near optimal) ways of obtaining a basic feasible solution. In the one we shall use, as many items as possible are sent by the cheapest routes. Hence we take

$$z_4 = 20 \text{ (making } x_4 = y_4 = 0)$$

$$y_3 = 45 \text{ (making } x_3 = z_3 = 0)$$

$$x_2 = 32 \text{ (making } y_2 = z_2 = 0)$$

and finally we must have

$$x_1 = 11, \ y_1 = z_1 = 10$$

We have six non-zero routes which correspond to our initial basic feasible solution. This can be written in tabular form (Table 11), the cost being shown in the bottom right-hand corner of each square.

### TABLE 11   THE INITIAL SOLUTION

| variables: | | x | y | z | totals |
|---|---|---|---|---|---|
| | 1 | 11 ($_{17}$) | 10 ($_{16}$) | 10 ($_{14}$) | 31 |
| | 2 | 32 ($_{11}$) | O ($_{14}$) | O ($_{13}$) | 32 |
| subscripts | 3 | O ($_{15}$) | 45 ($_{11}$) | O ($_{14}$) | 45 |
| | 4 | O ($_{12}$) | O ($_{11}$) | 20 ($_{10}$) | 20 |
| totals | | 43 | 55 | 30 | 128 |

The total cost for this solution is 1534 (using equation (c)).

Attempt problems 5.1.

### 5.2   CHANGE OF BASIS

We know that the optimal solution must be at a vertex, so one (inefficient) way of finding the optimum would be to calculate the cost at each vertex in turn and pick the smallest.

Suppose we move to the next vertex corresponding to $y_4$ being brought into the basis (i.e. being made non-zero). We come to the usual problem of deciding how much can be sent by this route. Suppose we denote by $\theta$ the maximum value of $y_4$ that can be introduced. Now in order to keep the row total at 20 we must reduce $z_4$ by $\theta$, and consequently to keep the correct column totals we must increase $z_1$ by $\theta$. The values of $x_1$ to $z_4$ must be non-negative, so that we must have $\theta \geqslant 0$, $20 - \theta \geqslant 0$, $10 + \theta \geqslant 0$, $10 - \theta \geqslant 0$ (Table 12). The largest value $\theta$ can take is 10 if all the constraints are to be satisfied. This means that $y_1$ becomes zero and hence leaves the basis to make way for $y_4$. The cost is now 1524, showing that we have

### TABLE 12   AN IMPROVED SOLUTION

| | | x | y | z | |
|---|---|---|---|---|---|
| | 1 | 11 ($_{17}$) | 10-$\theta$ ($_{16}$) | 10+$\theta$ ($_{14}$) | 31 |
| | 2 | 32 ($_{11}$) | O ($_{14}$) | O ($_{13}$) | 32 |
| | 3 | O ($_{15}$) | 45 ($_{11}$) | O ($_{14}$) | 45 |
| | 4 | O ($_{12}$) | $\theta$ ($_{11}$) | 20-$\theta$ ($_{10}$) | 20 |
| | | 43 | 55 | 30 | 128 |

indeed found a better solution - by chance!

Attempt problems 5.2.

5.3    THE COST FUNCTION

As with the standard simplex technique, we must try to rationalise our travels around the simplex by evaluating the cost function in terms of the zero variables (those not in the basis).    An inspection of the coefficients will then tell us which variables will reduce the cost if brought into the basis.

We will now return to  Table 11 and our initial solution in terms of the basis $x_1$, $y_1$, $z_1$, $x_2$, $y_3$, $z_4$.    If we solve equations (1) to (7) to find each of these variables in terms of the remainder, we can substitute the expressions into (c) to give the cost in terms of the zero variables. Doing this we eventually obtain

$$C = 1534 + x_3(15 - 11 + 16 - 17) + x_4(12 - 10 + 14 - 17) +$$
$$y_2(14 - 11 + 17 - 16) + y_4(11 - 10 + 14 - 16) +$$
$$z_2(13 - 11 + 17 - 14) + z_3(14 - 11 + 16 - 14) \tag{c1}$$
$$= 1534 + 3x_3 - x_4 + 4y_2 - y_4 + 5z_2 + 5z_3 \tag{c2}$$

which shows us that we can decrease the cost only by bring $x_4$ or $y_4$ into the basis.    Firstly, however, let us anlayse the structure of the coefficients as given in (c1).    We know that each coefficient *must* consist of the cost for the route considered, together with some costs associated with the basis variables.    The terms within each coefficient are seen to be of alternating signs, starting with the route cost itself.    If we join up the elements of the table which form any particular coefficient, we find they form a '*closed circuit*', starting and ending at the zero concerned. For example, if we denote basis variables by a ● and the zero in question by a *, we get the closed circuits of Fig.7.

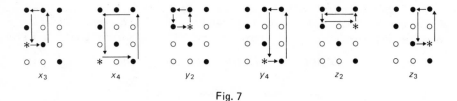

Fig. 7

We note that in such a closed circuit the only costs used are those at the right-angled bends.    In this particular tableau all the closed circuits are in the form of rectangles, but they can take more complicated forms (as does the one shown in Fig.8, which we shall be meeting later).

These closed circuits are of precisely the same form as those used when we changed the basis (Table 12).

Attempt problems 5.3.

Fig. 8

## 5.4 ALTERNATIVE OPTIMAL SOLUTIONS

Let us return to the problem: we know that the introduction of $x_4$ or $y_4$ will reduce the cost by 1 per unit transported. We have a choice, so let us take $y_4$ into the basis, as we did in Table 12. We now replace $\theta$ by 10 and have the situation as shown in Table 13.

TABLE 13  $y_4$ ENTERS THE BASIS

|   | x | y | z |   |
|---|---|---|---|---|
| 1 | 11<br>17 | O<br>16 | 20<br>14 | 31 |
| 2 | 32<br>11 | O<br>14 | O<br>13 | 32 |
| 3 | O<br>15 | 45<br>11 | O<br>14 | 45 |
| 4 | O<br>12 | 10<br>11 | 10<br>10 | 20 |
|   | 43 | 55 | 30 | 128 |

TABLE 14  $x_4$ TO ENTER THE BASIS

|   | x | y | z |   |
|---|---|---|---|---|
| 1 | 11−θ<br>17 | O<br>16 | 20+θ<br>14 | 31 |
| 2 | 32<br>11 | O<br>14 | O<br>13 | 32 |
| 3 | O<br>15 | 45<br>11 | O<br>14 | 45 |
| 4 | θ<br>12 | 10<br>11 | 10−θ<br>10 | 20 |
|   | 43 | 55 | 30 | 128 |

We must again evaluate the cost coefficients in terms of the zero variables to see whether we can improve upon the overall transportation cost. There is no need to evaluate the constant term at this stage, just the coefficients. As above we calculate them by following suitable (unique) closed circuits:

$$x_3 \; : \; 15 - 11 + 11 - 10 + 14 - 17 \; = \; 2$$
$$x_4 \; : \; 12 - 10 + 14 - 17 \qquad\qquad = -1$$
$$y_1 \; : \; 16 - 14 + 10 - 11 \qquad\qquad = \; 1$$
$$y_2 \; : \; 14 - 11 + 17 - 14 + 10 - 11 \; = \; 5$$
$$z_2 \; : \; 13 - 14 + 17 - 11 \qquad\qquad = \; 5$$
$$z_3 \; : \; 14 - 11 + 11 - 10 \qquad\qquad = \; 4$$

(See Fig.8 for closed circuit diagram for $y_2$.)

Hence only the introduction of $x_4$ will reduce the cost. How many items should be sent along this route?   Put $x_4 = \theta$, and following a closed circuit to keep the totals unchanged we must subtract $\theta$ from $z_4$, add $\theta$ to $z_1$, subtract $\theta$ from $x_1$: the only constraints affecting $\theta$ are $x_1$, $z_4$ which tell us that $11 - \theta \geqslant 0$ and $10 - \theta \geqslant 0$.   Therefore $\theta = 10$ is the largest allowable value (Table 14).

We now calculate the coefficients of the cost function again, in terms of the variables not in the basis:

$$x_3 \ : \ 15 - 11 + 11 - 12 \qquad\qquad = \ 3$$
$$y_1 \ : \ 16 - 11 + 12 - 17 \qquad\qquad = \ 0$$
$$y_2 \ : \ 14 - 11 + 12 - 11 \qquad\qquad = \ 4$$
$$z_2 \ : \ 13 - 14 + 17 - 11 \qquad\qquad = \ 5$$
$$z_3 \ : \ 14 - 14 + 17 - 12 + 11 - 11 = \ 5$$
$$z_4 \ : \ 10 - 14 + 17 - 12 \qquad\qquad = \ 1$$

Since none of these is negative we are unable to decrease the cost (from 1514), but the value of 0 for $y_1$ shows that there is an alternative solution involving that variable, with the same cost.   Adding $\theta$ to $y_1$ means subtracting $\theta$ from $x_1$, adding it to $x_4$ and subtracting it from $y_4$ (Table 15). The maximum value for such a $\theta$ is 1, leading to Table 16.

*TABLE* 15  $y_1$ *TO ENTER THE BASIS*      *TABLE* 16  *THE ALTERNATIVE SOLUTION*

|   | x | y | z |   |
|---|---|---|---|---|
| 1 | 1-θ / 17 | θ / 16 | 30 / 14 | 31 |
| 2 | 32 / 11 | 0 / 14 | 0 / 13 | 32 |
| 3 | 0 / 15 | 45 / 11 | 0 / 14 | 45 |
| 4 | 10+θ / 12 | 10-θ / 11 | 0 / 10 | 20 |
|   | 43 | 55 | 30 | 128 |

|   | x | y | z |   |
|---|---|---|---|---|
| 1 | 0 / 17 | 1 / 16 | 30 / 14 | 31 |
| 2 | 32 / 11 | 0 / 14 | 0 / 13 | 32 |
| 3 | 0 / 15 | 45 / 11 | 0 / 14 | 45 |
| 4 | 11 / 12 | 9 / 11 | 0 / 10 | 20 |
|   | 43 | 55 | 30 | 128 |

The new cost coefficients are

$$x_1 \ : \ 17 - 16 + 11 - 12 \qquad\qquad = \ 0$$
$$x_3 \ : \ 15 - 11 + 11 - 12 \qquad\qquad = \ 3$$
$$y_2 \ : \ 14 - 11 + 12 - 11 \qquad\qquad = \ 4$$
$$z_2 \ : \ 13 - 14 + 16 - 11 + 12 - 11 = \ 5$$
$$z_3 \ : \ 14 - 14 + 16 - 11 \qquad\qquad = \ 5$$
$$z_4 \ : \ 10 - 14 + 16 - 11 \qquad\qquad = \ 1$$

showing that the only alternative optimal solution is the one previously obtained.

It can be seen that the *cheapest* route is not used in either solution, but that the most expensive *is* in the first solution.

The important feature that makes this problem more amenable to hand solution is that it is self-correcting.  If a mistake is made in the calculation of the cost coefficients so that the wrong variable is brought into the basis, this will be rectified in later tableaux, since only the original information is used to calculate cost coefficients at each stage.

Attempt problems 5.4.

## 5.5  DANTZIG'S SOLUTION

Although the method discussed above is much shorter than the standard simplex tabular method, we will now examine an even shorter (and simpler) method of evaluating the cost coefficients, suggested by Dantzig.  Let us start our problem from the beginning again.  Associate with each row and column of the first tableau (Table 11) new variables (*shadow costs*), $r_1$, $r_2$, $r_3$, $r_4$ and $c_1$, $c_2$, $c_3$, respectively.  For any route *used* (i.e. in basis) we define the sum of the corresponding row and column shadow costs to be equal to the actual transport cost.  We then have

$$r_1 + c_1 = 17 \tag{8}$$

$$r_1 + c_2 = 16 \tag{9}$$

$$r_1 + c_3 = 14 \tag{10}$$

$$r_2 + c_1 = 11 \tag{11}$$

$$r_3 + c_2 = 11 \tag{12}$$

$$r_4 + c_3 = 10 \tag{13}$$

If we now calculate the cost coefficients in terms of these new variables we obtain

$$x_3 : \quad 15 - 11 + 16 - 17 = 15 - (r_3 + c_2) + (r_1 + c_1) - (r_1 + c_1)$$
$$= 15 - (r_3 + c_1)$$

$$x_4 : \quad 12 - (r_4 + c_3) + (r_1 + c_3) - (r_1 + c_1) \quad = 12 - (r_4 + c_1)$$

$$y_2 : \quad 14 - (r_1 + c_2) + (r_1 + c_1) - (r_2 + c_1) \quad = 14 - (r_2 + c_2)$$

$$y_4 : \quad 11 - (r_4 + c_3) + (r_1 + c_3) - (r_1 + c_2) \quad = 11 - (r_4 + c_2)$$

$$z_2 : \quad 13 - (r_2 + c_1) + (r_1 + c_1) - (r_1 + c_3) \quad = 13 - (r_2 + c_3)$$

$$z_3 : \quad 14 - (r_3 + c_2) + (r_1 + c_2) - (r_1 + c_3) \quad = 14 - (r_3 + c_3)$$

So in each case the cost coefficient equals the cost of that route minus the sum of its row and column shadow costs.  This relationship is quite general. However we still have to solve the six equations (8) to (13).  Fortunately there is always one more equation than there are variables when we have a (non-degenerate) basic feasible solution.  So, rather than have the non-determinacy that this implies, we can specify any value for *one* of the variables and this will then lead to a unique solution.

Let's take $r_1 = 0$ (say);  this leads to

$$c_1 = 17, \ c_2 = 16, \ c_3 = 14, \ r_2 = -6, \ r_3 = -5, \ r_4 = -4$$

(N.B. These variables are not required to be non-negative.)

We can now re-write Table 11 with an extra row and column to incorporate these variables (Table 11 **A**).

## TABLE 11A  THE INITIAL SOLUTION

|  | shadow costs | x 17 | y 16 | z 14 |  |
|---|---|---|---|---|---|
| 1 | 0 | 11 (17) | 10 (16) | 10 (14) | 31 |
| 2 | -6 | 32 (11) | [4] 0 (14) | [5] 0 (13) | 32 |
| 3 | -5 | [3] 0 (15) | 45 (11) | [5] 0 (14) | 45 |
| 4 | -4 | [-1] 0 (12) | [-1] 0 (11) | 20 (10) | 20 |
|  |  | 43 | 55 | 30 | 128 |

## TABLE 13A  THE SECOND SOLUTION

|  | shadow costs | x 17 | y 15 | z 14 |  |
|---|---|---|---|---|---|
| 1 | 0 | 11 (17) | [1] 0 (16) | 20 (14) | 31 |
| 2 | -6 | 32 (11) | [5] 0 (14) | [5] 0 (13) | 32 |
| 3 | -4 | [2] 0 (15) | 45 (11) | [4] 0 (14) | 45 |
| 4 | -4 | [-1] 0 (12) | 10 (11) | 10 (10) | 20 |
|  |  | 43 | 55 | 30 | 128 |

The cost coefficients can now be calculated by subtracting the row and column shadow costs from the actual cost involved - these are shown in the top left-hand corner of each square containing a zero element. They are, of course, the same as obtained previously. If we add $\theta$ to the new basic variable $y_4$, we find as before that the maximum allowable value for $\theta$ is 10. We start the next tableau (Table 13 A) by filling in the costs and basic feasible solution just obtained. If we assign the value 0 to $r_1$, we see that considering only those routes used, $c_1 = 17$ and $c_3 = 14$. It follows that $r_2 = -6$, $r_4 = -4$ and finally $c_2 = -15$ and $r_3 = -4$. We can now calculate the cost coefficients as before, and noting that $x_4$ has the only one that is negative, we add $\theta$, which must not be greater than 10, giving our final tableau (Table 14 A). Shadow costs, based on $r_1 = 0$, are shown, together with the cost coefficients, none of which is negative. The alternative solution can be obtained as before.

## TABLE 14A  AN OPTIMAL SOLUTION

|  | shadow costs | x 17 | y 16 | z 14 |  |
|---|---|---|---|---|---|
| 1 | 0 | 1 (17) | [0] 0 (16) | 10 (14) | 31 |
| 2 | -6 | 32 (11) | [4] 0 (14) | [5] 0 (13) | 32 |
| 3 | -5 | [3] 0 (15) | 45 (11) | [5] 0 (14) | 45 |
| 4 | -5 | 10 (12) | 10 (11) | [1] 0 (10) | 20 |
|  |  | 43 | 55 | 30 | 128 |

This small problem contains only 12 variables and 6 (or 7) constraints, but practical transportation problems often consist of hundreds of rows and columns. It is obvious that in such cases the use of the standard simplex tabular technique would be extremely inefficient and prohibitively costly in both time and storage even on the fastest computers. However, this neat method devised by Dantzig can be employed in hand calculations for even quite large problems, since it is self-correcting and needs only one tableau to be set up, provided previous costs and feasible solutions can be erased. Hence it is ideally suited to blackboard calculation.

Attempt problems 5.5.

## 5.6    DEGENERACY

As with all linear programming algorithms, the transportation technique relies on the use of non-degenerate basic feasible solutions. How does degeneracy manifest itself in the transportation problem? If you remember, we had one constraint less than the number of retailers (rows) plus depots (columns), and so any non-degenerate solution must contain this number of non-zero (basis) variables. Any number less than this constitutes a degenerate solution. Suppose that in our problem distributor $D_3$ has only 20 items and that retailer $R_1$ requires only 21. The initial tableau might well be as in Table 17, where as before we fill in the minimum-cost routes first. We see that there are only 5 non-zero variables instead of the $4 + 3 - 1 = 6$ required, hence this initial solution is degenerate. If we

*TABLE 17*
*AN INITIAL DEGENERATE SOLUTION*

|  | x | y | z |  |
|---|---|---|---|---|
| 1 | 11 _17_ | 10 _16_ | 0 _14_ | 21 |
| 2 | 32 _11_ | 0 _14_ | 0 _13_ | 32 |
| 3 | 0 _15_ | 45 _11_ | 0 _14_ | 45 |
| 4 | 0 _12_ | 0 _11_ | 20 _10_ | 20 |
|  | 43 | 55 | 20 | 118 |

*TABLE 18*
*COMBATTING DEGENERACY*

|  | x | y | z |  |
|---|---|---|---|---|
| 1 | 11+ε _17_ | 10−ε _16_ | 0 _14_ | 21 |
| 2 | 32 _11_ | 0 _14_ | 0 _13_ | 32 |
| 3 | 0 _15_ | 45 _11_ | 0 _14_ | 45 |
| 4 | 0 _12_ | 2ε _11_ | 20+ε _10_ | 20+3ε |
|  | 43+ε | 55+ε | 20+ε | 118+3ε |

try to continue as before, we find that there are not enough elements even to calculate the shadow costs. We have to make this solution non-degenerate in some way, the simplest being to modify some of the elements slightly as follows. Add ε (a small unknown positive constant) to each of the column totals and then add the total number of εs used to *just one* row, and, of course, to the overall total. Any new basic feasible solution is no longer degenerate (Table 18). We now continue as before, omitting the εs as soon as they are no longer needed to keep the solution non-degenerate.

All this procedure really does is pinpoint a route which, when brought into basis, breaks the degeneracy. If a degenerate solution occurs *during* calculations, it will be obvious which route will break the degeneracy, since it will have just been set to zero (fortuitously). To avoid errors it is often advisable to replace the offending zero by ε (and adjust corresponding totals) until the degeneracy is resolved. For example, consider a modified version of our problem. After the first tableau (Table 19) we find that two routes, $y_1$ and $z_4$, are reduced to zero by bringing $y_4$ into the basis. We can drop either $y_1$ or $z_4$ from the basis. Suppose we choose to drop $z_4$, then we must retain $y_1$, replacing its zero by $\varepsilon$ (Table 20).

### TABLE 19                    TABLE 20
### HANDLING INTERMEDIATE DEGENERACY

| | x | y | z | |
|---|---|---|---|---|
| 1 | 11 (17) | 10−θ (16) | 10+θ (14) | 31 |
| 2 | 32 (11) | O (14) | O (13) | 32 |
| 3 | O (15) | 45 (11) | O (14) | 45 |
| 4 | O (12) | O+θ (11) | 10−θ (10) | 10 |
| | 42 | 55 | 20 | 118 |

| | x | y | z | |
|---|---|---|---|---|
| 1 | 11 (17) | O+ε (16) | 20 (14) | 31+ε |
| 2 | 32 (11) | O (14) | O (13) | 32 |
| 3 | O (15) | 45 (11) | O (14) | 45 |
| 4 | O (12) | 10 (11) | O (10) | 10 |
| | 42 | 55+ε | 20 | 118+ε |

Attempt problems 5.6.

## 5.7   INEQUALITY

If the retailer requirements do not equal the available product stocks, then the incorporation of a *slack* depot or retailer with associated zero transport costs will change the problem into the form already discussed. For example, Tables 21 and 22 show us how to modify the problem if retailer $R_1$ really wants 41 or 21 items respectively.

### TABLE 21
### ORDERS NOT COMPLETELY FULFILLED

| | $D_1$ | $D_2$ | $D_3$ | $(D_4)$ | |
|---|---|---|---|---|---|
| $R_1$ | (17) | (16) | (14) | (0) | 41 |
| $R_2$ | (11) | (14) | (13) | (0) | 32 |
| $R_3$ | (15) | (11) | (14) | (0) | 45 |
| $R_4$ | (12) | (11) | (10) | (0) | 20 |
| | 43 | 55 | 30 | 10 | 138 |

### TABLE 22
### SOME STOCKS LEFT

| | $D_1$ | $D_2$ | $D_3$ | |
|---|---|---|---|---|
| $R_1$ | (17) | (16) | (14) | 21 |
| $R_2$ | (11) | (14) | (13) | 32 |
| $R_3$ | (15) | (11) | (14) | 45 |
| $R_4$ | (12) | (11) | (10) | 20 |
| $(R_5)$ | (0) | (0) | (0) | 10 |
| | 43 | 55 | 30 | 128 |

Attempt problems 5.7.

5.8   SUMMARY OF CHAPTER 5

Suppose there are r retailers and d depots (after the inclusion of any
necessary slacks).   Construct a table of size (d + 2) by (r + 2), with the
unit transport costs displayed in the central block of dxr squares.   The
requirements and stocks are then placed in the right and lower borders.
An initial basic feasible solution can be obtained by first using the
cheapest routes to their fullest extent and then filling in the final
squares as necessary.   Degeneracy (less than d + r - 1 non zero entries)
can be combatted by adding $\varepsilon$ to each of the d column totals, and d$\varepsilon$ to the
grand total and any one of the row totals.   The shadow costs are next
calculated and placed in the first row and column.   Set (any) one shadow
cost to zero and then evaluate the others by ensuring that for any route
*used*, the sum of its row and column shadow costs equals the unit transport
cost.

The cost coefficients are obtained by subtracting the row and column shadow
costs from the actual unit cost for the unused routes.   Any negative
coefficient shows that the corresponding variable may be brought into the
basis with advantage.   Add $\theta$ to this element and complete a 'closed
circuit', subtracting and adding $\theta$ to keep the totals unchanged.   The maximum value of $\theta$ is then chosen consistent with non-negativity constraints,
This leads to the next basic feasible solution.   Degeneracy at this stage
can be resolved by replacing the offending zero by $\varepsilon$ (and amending the
totals accordingly).

Attempt problems 5.8.

# Chapter 6

## 6.1 COMPETITION

In this chapter we shall be concerned with the formalisation in mathematical terms of the idea of competition.   In later chapters we shall learn how to solve a subset of such problems both graphically and by linear programming. This will, incidentally, give us an insight into a relationship between pairs of linear programming problems, called duality.

The simplest form of competition obviously occurs in parlour games such as chess, draughts, card games, etc. where the rules are precise.   However, competition exists also in everyday life, in business, politics, war, social rivalry, etc. but the 'rules' are often not quite so well defined.

Those forms of competition in which the rules are strictly formulated and known to the competitors we shall call *games*.   In order to consider every-day competition as games, we shall have to use simplifications and state precisely the 'rules' by which they are to be played.   A game is partly specified by the number of *players* - generally an n-person game in which the interests of the n players are in conflict.

For example in a game of rugby, although there are many players in the usual sense of the word, in game theory context we may, to a first order of approximation, assume there to be only 2 - the two teams - since each team consists of people with a common interest.

Attempt problems 6.1.

## 6.2 THE GOAL

In order to have a competition, there must be some goal (or goals) for the competitors to aim at; they must want to *win* something and in order to formulate the problem mathematically we must be able to give numerical values to the gains (*payoffs*) involved.   These normally take the form of monetary reward, but they can be thought of more rationally as *utilities*; for example, you might have a greater sense of elation in winning £5 from your greatest enemy than £10 from your best friend.

When the interests of the players are in direct conflict, i.e. what one loses another gains, the game is called *zero-sum* (no money (or utilities) created or destroyed, just a redistribution).   Unless we state otherwise, we shall in future be concerned only with 2 person, zero-sum games, i.e. games with two players, A and B, and A's losses being B's gains and vice versa.   For example, in dominoes, the person who loses might buy his opponent a drink - what the one person loses, the other gains.

Attempt problems 6.2.

## 6.3 STRATEGY

A game is made up of *moves*.   At every step each player will usually have to decide how to move, dependent on the previous moves (if known) of all players (a *personal* move). Sometimes, the moves can be *chance* moves, such as the outcome of tossing a penny. Theoretically, therefore, it would be possible *in advance of the game* for both players to write down their complete set of moves allowing for any combination of moves by their opponent(s) (provided we have a finite game). Complete strangers could then play the game for them without the original players even being present. Such a set of moves is called a *strategy*. A strategy need not be optimal in any sense.

Of course, a strategy normally consits of such a large number of possible moves that it is impractical to write it out in full - although it is theoretically possible.

      *Example 1:* (2 person, zero-sum game). Suppose a game consists of A tossing a coin and B guessing the outcome, heads or tails. B wins the game if he guesses correctly, otherwise he loses. B's strategy might be

      Always call heads

or Call heads and tails alternately

or If at any point of the game H heads and T tails have been obtained, call heads if H > T, tails if T > H

or Etc., etc.

Attempt problems 6.3.

6.4   OPTIMUM STRATEGY

Players A's *optimum strategy* is that strategy which guarantees him greatest gain (possibly over many games) no matter how the other players play.

      *Example 2:* Two players, A and B, each call out one of the numbers 1 and 2 simultaneously.

If they both call 1, no payment is made.

If they both call 2, B pays A 3p.

If A calls 1 and B calls 2, B pays A 1p.

But if A calls 2 and B calls 1, A pays B 1p.

On the face of it this game is not fair, since B might have to pay A 3p. We will now make a rational analysis of this game.

Firstly we tabulate all A's and B's possible strategies for each game and the the corresponding payoffs to A (the payoffs to B are just the negatives of these). The convention is to tabulate the payoffs to the player named at the left of the table.

*TABLE 23  THE PAYOFF MATRIX*

| B's strategies: | call 1 | call 2 |
|---|---|---|
| A's strategies: call 1 | 0 | 1 |
| call 2 | -1 | 3 |

This table is usually known as the *payoff matrix* and is assumed known to both players.

What are the optimum strategies for A and B?   A's reasoning is as follows: if I call 1, I can at worst draw with B (i.e. win 0); if I call 2, I can at worst lose 1p.   So to be on the safe side I will call 1 and guarantee that I will never lose anything.   Hence A can minimise his losses by calling 1. B's reasoning is as follows:  if I call 1, I can at worst draw with A; if I call 2, I can at worst lose 3p.   So to be on the safe side, I will call 1 and guarantee never to lose anything.   So B can minimise his losses by calling 1.   Hence both call 1 and draw, - 0 is the *value of the game* (to A), written V(A) = 0.

Let's check what happens if B calls 1 (his optimum) and A decides to change to 2.   A only loses by it, since now he must pay B 1p; similarly if B changes from his optimum while A calls 1, B can only lose by it.   It is for this reason that this game is said to be in *equilibrium* - it doesn't pay any one player to shift from his optimum.   (N.B. If they both shift *simultaneously*, then A obviously gains by it.)

Since the value of this game is 0, the game is said to be *fair*.

Attempt problems 6.4.

6.5   AN EXAMPLE

*Example 3:*  Two players, A and B, have the cards shown, (R = Red, B = Black).

A:   5R, 5B, 4B   (e.g. 5 of hearts, 5 of spades and 4 of clubs)

B:   5B, 3R, 1R

They place a total of 3 cards on the table face up in the following order: B first, then A, then B again.   As A places his card, a payoff is made and as B places his second card another payoff is made, each depending only on the last two cards played, as follows:  if both cards are of the same colour, A wins from B the difference between the numbers on the cards, otherwise B wins from A that difference.

What cards should A and B play?

Firstly in Table 24 we tabulate the strategies and payoffs to A (A's gains are B's losses), allowing for all possible outcomes.

### TABLE 24  THE FULL MATRIX

| B Plays | | 5B,3R | 5B,1R | 3R,1R | 3R,5B | 1R,5B | 1R,3R |
|---|---|---|---|---|---|---|---|
| A Plays | 5R: | 0+2=2 | 0+4=4 | 2+4=6 | 2+0=2 | 4+0=4 | 4+2=6 |
| | 5B: | 0−2=−2 | 0−4=−4 | −2−4=−6 | −2+0=−2 | −4+0=−4 | −4−2=−6 |
| | 4B: | 1−1=0 | 1−3=−2 | −1−3=−4 | −1+1=0 | −3+1=−2 | −3−1=−4 |

This shows that the order in which B plays his two cards is immaterial (obviously).   Hence B has only 3 really distinct strategies.   Table 24 can thus be reduced to Table 25, where the strategies have been named in the conventional manner.

B's reasoning is as follows:  if I play $B_1$, I can't lose more than 2, no matter how A plays; if I play $B_2$, I can't lose more than 4, no matter how A plays; if I play $B_3$, I can't lose more than 6, no matter how A plays. Therefore, no matter how A plays, I can limit my loss to 2 by playing $B_1$.

A's reasoning is as follows:  if I play $A_1$, I can win at least 2 no matter how B plays; if I play $A_2$, I can't lose more than 6; if I play $A_3$, I can't

TABLE 25  *THE REDUCED MATRIX*

|  | $B_1$ | $B_2$ | $B_3$ |
|---|---|---|---|
|  | (5B,3R) | (5B,1R) | (3R,1R) |
| $A_1$: 5R | 2 | 4 | 6 |
| $A_2$: 5B | -2 | -4 | -6 |
| $A_3$: 4B | O | -2 | -4 |

lose more than 4.  Hence, no matter how B plays, I can ensure a gain of 2 by playing $A_1$.

Since A expects to gain 2, and B expects to lose 2, $V(A) = 2$, the optimum strategies being $(A_1,B_1)$.  This game is not fair, but it is in equilibrium.

Attempt problems 6.5.

## 6.6  MAXIMIN AND MINIMAX

Perhaps a method of obtaining the solutions to these games in a more compact form is now becoming clear.

Consider example 2 (page 35) again, but with the payoff matrix extended as shown in Table 26.

TABLE 26  *MINIMAX AND MAXIMIN*

| B's strategies: | Shout 1 | Shout 2 | Row min. |
|---|---|---|---|
| A's strategies  Shout 1 | O | 1 | O |
| Shout 2 | -1 | 3 | -1 |
| Col max. | O | 3 | |

For A's optimum strategy we first look at the maximum he can lose, or rather the minimum he can guarantee to win for each of his strategies (since this is a table of A's gains).  This corresponds to the minimum element in each row of the table (see last column).  Then we determine the most favourable of these, i.e. the maximum value - in this table it is O - which is called the *maximin* value of the game (it is the *maximum* of the row *mini*ma).

For B's optimum strategy we do the same, remembering that B's payoff matrix is the *negative* of A's.  Hence instead of looking for maxima we look for minima and vice versa (see last row).  The most favourable of these to B is the minimum, called the *minimax* value of the game (the *minimum* of the column *maxi*ma).  Here the maximin and minimax values of the game are equal and in turn are equal to the value of the game, which is a *saddle point* of the payoff matrix (i.e. is at the same time the minimum of its row and the maximum of its column).

Attempt problems 6.6.

## 6.7  FULL INFORMATION

*Example 4:*  The game is the same as in example 3 (p. 36), except that all cards are placed on the table simultaneously.  The payoff matrix, the reasoning and the solution will be as before.  However, there is a difference between these two games; the first (example 3) is one of *full information*, i.e. at each stage of the game all previous 'moves' are known

to both players, whilst example 4 is not, since both moves are made
simultaneously, i.e. each player does not know his opponent's latest move.
It can be shown that the first type always has a saddle point and hence a
solution, whereas this is not necessarily true of the second.   Examples of
type 1 games are noughts and crosses, draughts and chess; examples of type 2
are bridge, poker, etc.

Attempt problems 6.7.

6.8   PURE AND MIXED STRATEGIES

   *Example 5:*   Players A and B simultaneously call out either of the
numbers 1 and 2.   If their sum is even, B pays A that number of £1, if odd
A pays B.   How should A and B play?   This is a game without full
information.

*TABLE* 27

| B:<br>A | 1 | 2 | Row<br>Min |
|---|---|---|---|
| 1 | 2 | -3 | -3 |
| 2 | -3 | 4 | -3 |
| Col max | 2 | 4 | |

For A's optimum strategy we have the maximin value of the game, which, from
Table 27, is -3 (from calling either 1 or 2).   For B's optimum strategy we
have the minimax value of the game, +2 (from calling 1), i.e. A expects to
lose 3 and B expects to lose 2.   However, since this is a zero-sum game
they *both* can't lose!   This apparent contradiction has arisen because the
maximin and minimax values are not equal - there is no saddle point to the
game.   In fact there is no predictable solution to a single game as in the
previous examples.   We bypass this by considering the outcome over many
plays of the game.   The reasoning runs as follows:  if A calls 1 repeatedly
and B calls 2 repeatedly, A will obviously change to 2 and B will then
change to 1, etc., i.e. we have *instability* (a consequence of there being no
saddle point to the game).   The only way out of this instability is for a
player not to play the same strategy continually, so that his opponent
cannot change his strategy beneficially.   The players must therefore play
their strategies in a random manner, the frequencies being chosen to give
them their 'best' payoffs over many games.   We say that the players have
moved from *pure* strategies to *mixed* strategies.

It will be shown later that if A calls 1 and 2 in the ratio 7 to 5, and if B
does the same, B will end up by winning £1/12 on average, i.e. A will have
cut his expected loss from £3 to £1/12.

As we have seen, this game has no solution in pure strategies, but it does
have one in mixed strategies.

Attempt problems 6.8.

6.9   REDUCTION OF THE GAME

It is usual to try to reduce the size of a game in order to solve it more
easily.   There are two basic ways in which this may be accomplished.   We
eliminate:

1)   Duplicate strategies (as in example 3, page 36)
2)   Dominated strategies - those where the payoffs to A in any particular

row are all less than or equal to those in another row, or those where the
payoffs to A in a particular column are all greater than or equal to those
in another column.  In either case the eliminated strategies would never be
used since they could not be optimal.

*Example 6:*  Suppose we have the following payoff matrix to A.

### TABLE 28  REDUNDANT STRATEGIES

| B:<br>A | $B_1$ | $B_2$ | $B_3$ | $B_4$ |
|---|---|---|---|---|
| $A_1$: | 0 | -1 | 2 | -4 |
| $A_2$: | 1 | 3 | 3 | 6 |
| $A_3$: | 2 | -4 | 5 | 1 |

Strategy $A_1$ is redundant, since it is dominated by $A_2$, i.e. A will never
play $A_1$ (rationally), since he can always gain more by playing $A_2$.
Similarly $B_3$ is redundant through being dominated by $B_2$, i.e. B will always
gain more by playing $B_2$ rather than $B_3$.   Hence the game reduces to that of
Table 29, which in turn reduces to that shown in Table 30, since $B_2$ now

### TABLE 29  THE REDUCED MATRIX

| B:<br>A | $B_1$ | $B_2$ | $B_4$ |
|---|---|---|---|
| $A_2$: | 1 | 3 | 6 |
| $A_3$: | 2 | -4 | 1 |

### TABLE 30  THE FINAL SIMPLIFICATION

| B:<br>A | $B_1$ | $B_2$ |
|---|---|---|
| $A_2$: | 1 | 3 |
| $A_3$: | 2 | -4 |

dominates $B_4$.   The game is now of a more manageable size.   We shall be
discussing how to solve games with no saddle point, such as this one, in the
next chapter.

Attempt problems 6.9.

6.10   SUMMARY OF CHAPTER 6

We classify *games* by the following:

1)   Number of players

2)   Number of moves

3)   Whether or not they are zero-sum

4)   Whether or not they are of full information

and *moves* by 'personal' or 'chance'.   Zero-sum games with full information
have a saddle point and hence have a solution using pure strategies.   Zero-
sum games without full information can often be 'solved' on a frequency
basis only, by using mixed strategies.

# Chapter 7

## 7.1    SOLUTION OF GAMES IN MIXED STRATEGIES

We now list three results which are the basis for all the work in this chapter.

*Result 1.*    It can be proved that by using mixed strategies every finite, two-person, zero-sum game has a solution, this solution being at the same time best for both players, and the common value being the value of the game.    The solution also has the equilibrium property of the saddle game, in that any deviation from these mixed strategies by one player cannot increase his profit (over many games).

*Result 2.*    It is easily shown that this value of the game, $V(A)$, lies between the maximin ($\alpha$) and minimax ($\beta$) values of the game.    This follows, since from A's point of view the value of the game is A's optimum, and hence must be better than (or equal to) his best pure strategy (maximin), i.e. $V(A) \geqslant \alpha$.    Similarly, from B's point of view the value of the game must be better than or equal to the minimax strategy, i.e. $- V(A) \geqslant -\beta$, giving $\alpha \leqslant V(A) \leqslant \beta$.

Attempt problems 7.1.

## 7.2    SOLUTION OF GAMES IN MIXED STRATEGIES (continued)

*Result 3.*    Generally, it will be the case that not all of A's moves appear in his optimal mixed strategy.    Let $S_A^*$ consist of those which do appear, and $S_B^*$ consist of those which B actually uses, then the following important result holds.

If one player (say A) adheres to his optimum strategy, then no matter what proportions of the moves in B's optimum strategy $S_B^*$ are used, the value of the game remains unaltered.    (N.B. If B uses moves *not* in his optimal strategy, then he may well change the value of the game to his own disadvantage!)

*Proof* (We 'prove' the result for a specific case, but the working is easily generalised.)

Suppose    $S_A^* = \left\{ \begin{matrix} A_1, & A_4, & A_7 \\ p_1, & p_4, & p_7 \end{matrix} \right\}$    $S_B^* = \left\{ \begin{matrix} B_3, & B_6, & B_7, & B_9 \\ q_3, & q_6, & q_7, & q_9 \end{matrix} \right\}$

(say) where A uses $A_1$ a proportion $p_1$ of the time, etc.    For convenience we make all the elements of the payoff matrix positive by the addition of a suitable constant ($V(A)$ will just be increased by this constant).

Then

$$p_1 + p_4 + p_7 = 1 \qquad\qquad q_3 + q_6 + q_7 + q_9 = 1$$
$$p_1, \ \ p_4, \ \ p_7 > 0 \qquad\qquad q_3, \ \ q_6, \ \ q_7, \ \ q_9 > 0 \tag{1}$$

Let V be the value of the game (to A), and suppose A adheres to his optimal mixed strategy $S_A^*$.   If $V_3$ is the value of the game (to A) when B uses $B_3$ only (etc.), then

$$V = q_3 V_3 + q_6 V_6 + q_7 V_7 + q_9 V_9 \qquad\qquad (2)$$

Now $V_3 \geqslant V$, $V_6 \geqslant V$, $V_7 \geqslant V$, $V_9 \geqslant V$, since deviation from B's optimum can only increase the value of the game to A.

Suppose $V_3 > V$ (say) then, from (2)

$$V = q_3 V_3 + q_6 V_6 + q_7 V_7 + q_9 V_9 > V(q_3 + q_6 + q_7 + q_9) = V \text{ using (1)}$$

This is a contradiction, so $V_3 = V$.   Similarly $V_6 = V$, $V_7 = V$, $V_9 = V$. The corresponding result when A moves from his optimal strategy can be similarly obtained.

Attempt problems 7.2.

## 7.3   SOLUTION WHEN THERE IS NO SADDLE POINT

The following example will explain the method.

   *Example 8:*  (cf. example 5, page 38).   Consider the game with the following payoff matrix (Table 31).

### TABLE 31

| B: <br> A | $B_1$ | $B_2$ | min |
|---|---|---|---|
| $A_1$: | 2 | -3 | $\boxed{-3}$ |
| $A_2$: | -3 | 4 | $\boxed{-3}$ |
| max | $\boxed{2}$ | 4 | |

As noted previously this game has no saddle point.

   $\alpha$ = maximin = -3

   $\beta$ = minimax = +2 .

Hence V(A) lies between -3 and +2.

Since neither A nor B has a pure strategy as optimum, all strategies will have to be used.   (If the game were of a larger size, we could not make such an assumption.)   If

$$S_A^* = \left\{ \begin{array}{cc} A_1 & A_2 \\ p_1 & p_2 \end{array} \right\} \qquad S_B^* = \left\{ \begin{array}{cc} B_1 & B_2 \\ q_1 & q_2 \end{array} \right\}$$

Then by Result 3, if A keeps to $S_A^*$, V(A) does not depend on the frequencies with which $B_1$, $B_2$ are used.   Hence

$$V = p_1(2) + p_2(-3) \qquad \text{(if B uses } B_1 \text{ only)}$$

and   $V = p_1(-3) + p_2(4) \qquad \text{(if B uses } B_2 \text{ only)}$

Equating these we get

   $p_1 = 7/12$, $p_2 = 5/12$, (using $p_1 + p_2 = 1$) and $V = -1/12$

Similarly, if B keeps to $S_B^*$, then

$$V = q_1(2) + q_2(-3) \qquad \text{(A uses } A_1),$$

and   $V = q_1(-3) + q_2(4) \qquad \text{(A uses } A_2),$

giving $q_1 = 7/12$, $q = 5/12$ and $V = -1/12$ as before.

Hence

$$S_A^* = \left\{ \begin{matrix} A_1 & A_2 \\ 7/12 & 5/12 \end{matrix} \right\} \qquad S_B^* = \left\{ \begin{matrix} B_1 & B_2 \\ 7/12 & 5/12 \end{matrix} \right\}$$

and the value of the game is - £1/12 to A, i.e. B will win £1/12 on average. Attempt problems 7.3.

**7.4    GEOMETRIC SOLUTION**

Consider the general 2 x 2 game with matrix of Table 32.

### TABLE 32   THE GENERAL 2 x 2 PAYOFF MATRIX

| B:<br>A | $B_1$ | $B_2$ |
|---|---|---|
| $A_1$: | $a_{11}$ | $a_{12}$ |
| $A_2$: | $a_{21}$ | $a_{22}$ |

Let us suppose it has no saddle point, since that case is now trivial.

If B just plays $B_1$, and A plays $A_1$ and $A_2$ in proportions $p_1$, $p_2$, (not necessarily optimum), then A will receive, on average,

$$V = a_{11} p_1 + a_{21} p_2 \qquad \text{where } p_1 + p_2 = 1$$

Let us now draw this result (Fig. 9), where we are treating the a's as fixed, but the p's as unknowns.

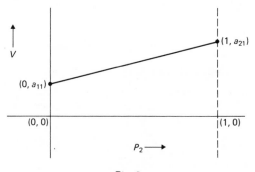

Fig. 9

We plot V against $p_2$ and find it to be a straight line cutting the line $p_2 = 0$ at $a_{11}$ and $p_2 = 1$ at $a_{21}$.   We are now in a position to find the optimal strategies.   If

$$S_A^* = \left\{ \begin{matrix} A_1 & A_2 \\ p_1 & p_2 \end{matrix} \right\} \qquad S_B^* = \left\{ \begin{matrix} B_1 & B_2 \\ q_1 & q_2 \end{matrix} \right\}$$

then the simultaneous equations to be solved for $p_1$, $p_2$, V are

$$V = a_{11} p_1 + a_{21} p_2 \qquad \text{(B plays } B_1\text{)}$$
$$V = a_{12} p_1 + a_{22} p_2 \qquad \text{(B plays } B_2\text{)}$$
$$(p_1 + p_2 = 1)$$

We plot both of these lines (Fig. 10) and since the equations are to be solved simultaneously, their intersection will correspond to the solution: $p_2 = OP$, $p_1 = PI = 1 - OP$, $V = PM$

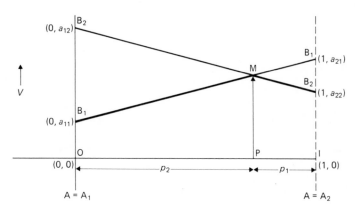

Fig. 10

There is, however, another way to look at the graph. The intercepts on the line $p_2 = 0$, i.e. $(a_{11}, a_{12})$, correspond to the values in $A_1$, and those on $p_2 = 1$, $(a_{21}, a_{22})$, correspond to those in $A_2$. The line $B_2, B_2$ joins the values in $B_2$, and $B_1, B_1$ those in $B_1$. Now, A is free to choose $p_2$ as he wishes and B to choose which strategy he uses, so if A chooses any $p_2$ less than p, B will counter by choosing $B_1$, since he will lose less. Similarly, if A chooses $p_2$ greater than p, B will play $B_2$ (i.e. B moves on the thick line). Hence A will choose $p_2 = p$, since this will guarantee him most. A similar line of argument will give us B's optimum strategy from Fig. 11 (note the relabelling of the axes).

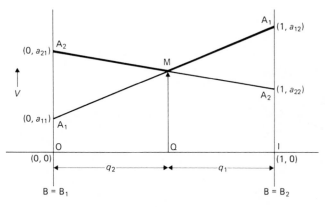

Fig. 11

We shall now look at this method on an actual numerical problem.

*Example 9:* (Payoff matrix of Table 33 is as for example 8).

*TABLE* 33

|        | B₁  | B₂  |
|--------|-----|-----|
| A₁:    | 2   | −3  |
| A₂:    | −3  | 4   |

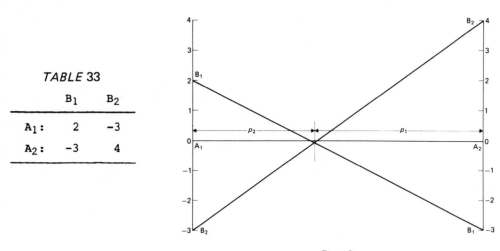

Fig. 12

By careful measurement in Fig. 12 we can confirm the results obtained previously, namely that $p_1 = 7/12$, $p_2 = 5/12$, $V = -1/12$.

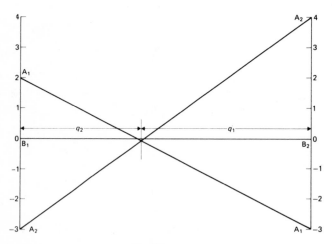

Fig. 13

Similarly from Fig. 13 $q_1 = 7/12$, $q_2 = 5/12$, $V = -1/12$.

Attempt problems 7.4.

7.5 GENERALISATION TO A 2xN GAME

Suppose we have a 2x3 game with no saddle point. We know that A must use both of his strategies, but we are not sure whether B uses all 3 or just 2 of his strategies.     We can easily see that, in general, B will not use more strategies than A, since we would obtain 3 equations in only 2 unknowns, $p_1$ and $p_2$.

Use of the geometric method together with Result 3 will enable us to solve this problem.

Consider the problem shown in Table 34.

### TABLE 34   A 2 x 3 GAME

|        | $B_1$ | $B_2$ | $B_3$ |
|--------|-------|-------|-------|
| $A_1$: | -1    | 2     | 4     |
| $A_2$: | 4     | 3     | 0     |

There is no saddle point, so we proceed as in the 2x2 case, but include the line $B_3$ $B_3$ (Fig. 14).

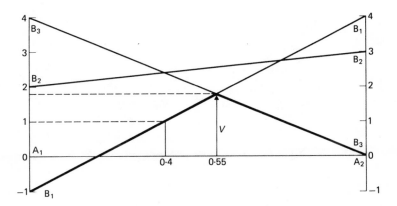

Fig. 14

By suitable choices of strategy, B can force the value of the game to values on the thick line.   For example, if A played $A_1$ only, the value of the game would be -1, since B would play $B_1$ only.    Similarly, if A played $A_1$ 0.6 of the time and $A_2$ 0.4 of the time, then B would still play $B_1$ and the value of the game would be 1.    Hence A will play his most favourable strategy under these circumstances, namely $p_1 = 0.45$, $p_2 = 0.55$, when the value of the game is about 1.8 (compare the graphical solution of a linear programming problem).    B will obviously play both $B_1$ and $B_3$ in this case (and not $B_2$). Knowing this, we can now solve graphically for $q_1$, $q_3$, V, since the problem is now reduced to a 2x2 game.   We can obviously extend this method from N = 3 to the general case.

Attempt problems 7.5.

7.6    SUMMARY OF CHAPTER 7

2xN games are solved as follows:

1)     Check for dominance.

2)     Check for saddle.

3)     If there is no saddle, reduce to 2x2 game by geometric approach.

4)     Solve algebraically (for accuracy) or graphically (for speed).

N.B. Don't try to remember the formula obtained for the general solution of a 2x2 game, just the method.

Attempt problems 7.6.

# Chapter 8

## 8.1 THE GENERAL SOLUTION OF AN mxn ZERO-SUM GAME

Consider the general mxn zero-sum game : A has moves $A_1$ ... $A_m$: B has moves $B_1$ ... $B_n$. $S_A^*$, $S_B^*$ will consist of some of these $A_i$'s and $B_j$'s. We can write the general payoff matrix $a$ to A as in Table 35.

### TABLE 35 THE GENERAL GAME

| B : | $B_1$ | $B_2$ | ... | $B_j$ | ... | $B_n$ |
|-----|-------|-------|-----|-------|-----|-------|
| A   |       |       |     |       |     |       |
| A : | $a_{11}$ | $a_{12}$ ... | | $a_{1j}$ ... | | $a_{1n}$ |
| A : | $a_{21}$ | $a_{22}$ ... | | $a_{2j}$ ... | | $a_{2n}$ |
| .   |       |       |     |       |     |       |
| .   | .     | .     |     | .     |     | .     |
| .   | .     | .     |     | .     |     | .     |
| $A_i$ : | $a_{i1}$ | $a_{i2}$ ... | | $a_{ij}$ ... | | $a_{in}$ |
| .   |       |       |     |       |     |       |
| .   | .     | .     |     | .     |     | .     |
| .   | .     | .     |     | .     |     | .     |
| $A_m$ : | $a_{m1}$ | $a_{m2}$ ... | | $a_{mj}$ ... | | $a_{mn}$ |

Let $S_A^! = S_A^*$, A's optimum mixed strategy, except that *all* the moves of A are included, those not occurring in $S_A^*$ having $p_i = 0$. Then, defining $S_B^*$ similarly we have

$$S_A^! = \begin{Bmatrix} A_1 & A_2 .... & A_m \\ p_1 & p_2 .... & p_m \end{Bmatrix} \text{ and } S_B^! = \begin{Bmatrix} B_1 & B_2 .... & B_n \\ q_1 & q_2 .... & q_n \end{Bmatrix}$$

where

$$p_1 + p_2 + ... + p_m = 1 \qquad\qquad q_1 + q_2 + ... + q_n = 1 \qquad (1)$$

$$p_i \geqslant 0 \ (i = 1 ... m) \qquad\qquad q_j \geqslant 0 \ (j = 1 ... n)$$

we must assume that the value of the game to A is positive (we can add a constant to each element of $a$) and that the value of the game is now V to A $(V \geqslant 0)$.

As $S_A^!$ is A's optimum strategy, then the average payoff corresponding to B using $B_j$ only is $p_1 a_{1j} + p_2 a_{2j} + .... + p_m a_{mj}$, which is not less than V, since $B_j$ does not necessarily belong to $S_B^*$, i.e.

$$p_1 a_{1j} + p_2 a_{2j} + .... + p_m a_{mj} \geqslant V \qquad (2)$$

This relationship must be true for all j (j = 1 ... n).

Put $x_1 = p_1/V$, $x_2 = p_2/V$, etc. then (2) becomes

$$x_1\, a_{1j} + x_2\, a_{2j} + \ldots + x_m\, a_{mj} \geqslant 1 \tag{2a}$$

for j = 1 ... n

Equation (1) becomes

$$x_1 + x_2 + \ldots + x_m = 1/V \tag{1a}$$

where $x_i \geqslant 0$ (i = 1 ... m).   Since A wishes to make $\overline{V}$ as large as possible, this is the same as minimising 1/V.

Our problem can therefore be written as

$$\text{minimise} \quad (1/V) = x_1 + x_2 + \ldots + x_m$$

subject to the n inequalities (constraints)

$$x_1\, a_{11} + x_2\, a_{21} + \ldots + x_m\, a_{m1} \geqslant 1$$

$$x_1\, a_{12} + x_2\, a_{22} \qquad\quad + x_m\, a_{m2} \geqslant 1$$

.

.

.

.

.

.

.

$$x_1\, a_{1n} + x_2\, a_{2n} + \ldots + x_m\, a_{mn} \geqslant 1$$

$$x_1 \geqslant 0 \ldots\ldots x_m \geqslant 0$$

(3)

which is a standard linear programming problem.

Attempt problems 8.1

**8.2   THE DUAL PROBLEM**

Looking at the problem from B's point of view, the value of the game is −V and, as before,

$$q_1\, b_{1j} + q_2\, b_{2j} + \ldots + q_n\, b_{nj} \geqslant -V \tag{4}$$

for j = 1 ... m

where the b's are the elements of B's payoff matrix ($b_{ij} = -a_{ji}$), and

$$q_1 + q_2 + \ldots + q_n = 1. \tag{from (1)}$$

Putting $y_1 = q_1/V$ etc., we obtain

$$y_1\, b_{1j} + y_2\, b_{2j} + \ldots + y_n\, b_{nj} \geqslant -1 \tag{4a}$$

j = 1 ... m

subject to

$$y_1 + y_2 + \ldots + y_n = \frac{1}{V}, \; y_j \geqslant 0 \quad (j = 1 \ldots m) \tag{1b}$$

We can now write (4a) in terms of the elements of $a$.

$$-y_1\, a_{j1} - y_2\, a_{j2} \ldots -y\ a_{jn} \geqslant -1 \qquad (j = 1 \ldots m)$$

or    $$y_1\, a_{ji} + y_2\, a_{j2} + \ldots + y_n\, a_{jn} \leqslant 1 \qquad (j = 1 \ldots m) \tag{4b}$$

We must note the change in direction of the inequality caused by multiplying equation (4a) by -1.

B wishes to make -V as large as possible, i.e.V as small as possible or 1/V as large as possible.   Hence B's problem is

$$\text{maximise} \quad (1/V) = y_1 + y_2 + \ldots + y_n$$

subject to

$$y_1 a_{11} + y_2 a_{12} \ldots\ldots + y_n a_{1n} \leqslant 1$$
$$y_1 a_{21} + y_2 a_{22} \qquad + y_n a_{2n} \leqslant 1$$

.
.
.

$$y_1 a_{m1} + y_2 a_{m2} \qquad + y_n a_{mn} \leqslant 1$$
$$y_1 \geqslant 0 \ldots \qquad\qquad y_n \geqslant 0$$

(5)

which is also a standard linear programming problem.

Using Result 1 of Chapter 7, we see that the optimal solutions to these two linear programming problems (3) and (5) must be such that they both give the same value for 1/V,  hence they are said to be *duals*.

Attempt problems 8.2

8.3   THE DUAL PROBLEM (continued)

*Example* 10:   Consider the game of Table 36.

### TABLE 36

| B : | $B_1$ | $B_2$ | $B_3$ |
|---|---|---|---|
| A | | | |
| $A_1$: | -1 | 0 | 1 |
| $A_2$: | 3 | 2 | -1 |
| $A_3$: | -3 | 1 | 0 |

We can easily see that this game has no saddle or dominance, hence it cannot be solved by graphical methods.   Firstly we must ensure that V is positive, which is most easily done by adding 4 to all terms (Table 37),

### TABLE 37   THE POSITIVE MATRIX

| B : | $B_1$ | $B_2$ | $B_3$ |
|---|---|---|---|
| A | | | |
| $A_1$: | 3 | 4 | 5 |
| $A_2$: | 7 | 6 | 3 |
| $A_3$: | 1 | 5 | 4 |

and turn it into the following linear programming problem.

$$\text{minimise} \quad \frac{1}{V} = x_1 + x_2 + x_3$$

subject to the constraints

$$3x_1 + 7x_2 + x_3 \geqslant 1$$
$$4x_1 + 6x_2 + 5x_3 \geqslant 1$$

$$5x_1 + 3x_2 + 4x_3 \geqslant 1$$

$$x_1, x_2, x_3 \geqslant 0$$

and obtain the tableau of Table 38, in which the original equation $P + x_1 + x_2 + x_3 + M(u + v + w) = 0$ has been put into standard form by subtracting from it M times each constraint.

### TABLE 38  THE INITIAL TABLEAU

|   | $x_1$ | $x_2$ | $x_3$ | r | s | t | u | v | w |   |   | Check |
|---|-------|-------|-------|---|---|---|---|---|---|---|---|-------|
| P | 1–12M | ⟨1–16M⟩ | 1–10M | M | M | M | 0 | 0 | 0 | –3M |   | 3–38M |
| u | 3 | [7] | 1 | –1 | 0 | 0 | 1 | 0 | 0 | 1 | ⟨1/7⟩ | 12 |
| v | 4 | 6 | 5 | 0 | –1 | 0 | 0 | 1 | 0 | 1 | 1/6 | 16 |
| w | 5 | 3 | 4 | 0 | 0 | –1 | 0 | 0 | 1 | 1 | 1/3 | 13 |

By the standard simplex technique we obtain the solution

$$x_1 = {}^2/13, \; x_2 = {}^1/13, \; x_3 = 0 \quad (s = {}^1/13),$$

hence

$$P = 1/V = {}^3/13 \text{ giving } V = {}^{13}/3$$

$$p_1 = {}^2/13, \; p_2 = {}^1/13, \; p_3 = 0.$$

The true value of the game is then

$$V = {}^{13}/3 - 4 = {}^1/3 \text{ to A.}$$

Attempt problems 8.3.

8.4    DUALITY

We have seen from the last section that for any linear programming problem associated with a matrix game there is a dual problem which must have the same optimal solution (value for V).    There is, however, a much stronger result (the proof being beyond the mathematical skills assumed here) which states that for *any* linear programming problem there exists a dual and that if. the (optimal) solution to one is finite then so is the other and they have the same value (of their profit functions) at their optima.    Further-more, having solved a given problem by the tabular simplex method, the relationship between the problem and dual is so strong that we can also deduce the values of its variables at that optimum.

The method of obtaining the dual from a given problem is as follows.    If the problem is a maximisation then the constraints (other than the usual positivity of the variables) must be of the form "$\leqslant$", if it is a minimisation then they must all be "$\geqslant$".    The following are then interchanged

$$\text{maximisation} \; : \; \text{minimisation}$$

$$\leqslant \; : \; \geqslant$$

$$\text{P-coefficients} \; : \; \text{Constant terms}$$

$$\text{Column coefficients} \; : \; \text{Row coefficients}$$

The following worked examples will help to fix ideas and demonstrate the results.

*Example* 11:   To find the dual of:

Maximise    $P = 3x_1 + 4x_2 - x_3$

subject to  $x_1 + x_2 + x_3 \geqslant 4$

$2x_1 - 4x_2 - x_3 \leqslant 10$

$x_1 + 2x_2 + 5x_3 = 8$

$x_1, x_2, x_3 \geqslant 0$

All the constraints must be "$\leqslant$" for a maximisation problem, and so we replace the first by

$-x_1 - x_2 - x_2 \leqslant -4$

and the third by the *pair* of inequalities

$x_1 + 2x_2 + 5x_3 \leqslant 8$

$-x_1 - 2x_2 - 5x_3 \leqslant -8$

The problem then becomes

maximise $P = 3x_1 + 4x_2 - x_3$

subject to  $-x_1 - x_2 - x_3 \leqslant -4$

$2x_1 - 4x_2 - x_3 \leqslant 10$

$x_1 + 2x_2 + 5x_3 \leqslant 8$

$-x_1 - 2x_2 - 5x_3 \leqslant -8$

$x_1, x_2, x_3 \geqslant 0$

The dual problem is

minimise $Q = -4y_1 + 10y_2 + 8y_3 - 8y_4$

subject to  $-y_1 + 2y_2 + y_3 - y_4 \geqslant 3$

$-y_1 - 4y_2 + 2y_3 - 2y_4 \geqslant 4$

$-y_1 - y_2 + 5y_3 - 5y_4 \geqslant -1$

$y_1, y_2, y_3, y_4 \geqslant 0$

*Example 12:*  The following two problems are duals.

(a) Maximise  $10x_1 + 15x_2$

subject to  $x_1 + x_2 \leqslant 50$

$2x_1 + x_2 \leqslant 110$

$x_1 + 2x_2 \leqslant 80$

$x_1 + 5x_2 \leqslant 185$

$5x_1 + 6x_2 \leqslant 300$

$x_1, x_2 \geqslant 0$

(b) Minimise  $50y_1 + 110y_2 + 80y_3 + 185y_4 + 300y_5$

subject to  $y_1 + 2y_2 + y_3 + y_4 + 5y_5 \geqslant 10$

$y_1 + y_2 + 2y_3 + 5y_4 + 6y_5 \geqslant 15$

$y_1, \ldots, y_5 \geqslant 0$

The final tableaux of each of these problems are shown in Tables 39 and 40 (after any dummy variables have been eliminated).

### TABLE 39

|     | $x_1$ | $x_2$ | $u_1$ | $u_2$ | $u_3$ | $u_4$ | $u_5$ |     |
|-----|-----|-----|-----|-----|-----|-----|-----|-----|
| P   | 0   | 0   | 5   | 0   | 5   | 0   | 0   | 650 |
| $u_4$ | 0 | 0   | 3   | 0   | -4  | 1   | 0   | 15  |
| $u_2$ | 0 | 0   | -3  | 1   | 1   | 0   | 0   | 40  |
| $x_1$ | 1 | 0   | 2   | 0   | -1  | 0   | 0   | 20  |
| $x_2$ | 0 | 1   | -1  | 0   | 1   | 0   | 0   | 30  |
| $u_5$ | 0 | 0   | -4  | 0   | -1  | 0   | 1   | 20  |

### TABLE 40

|     | $y_1$ | $y_2$ | $y_3$ | $y_4$ | $y_5$ | $r_1$ | $r_2$ |      |
|-----|-----|-----|-----|-----|-----|-----|-----|------|
| P   | 0   | 40  | 0   | 15  | 20  | 20  | 30  | -650 |
| $y_1$ | 1 | 3   | 0   | -3  | 4   | -2  | 1   | 5    |
| $y_3$ | 0 | -1  | 1   | 4   | 1   | 1   | -1  | 5    |

As we expected, 650 is the optimal value of both problems. However, if we associate the slacks of one problem with the true variables of the other, we see that the coefficients in the P equation of the one problem give the solutions to the other, as shown in Table 41.

### TABLE 41 THE RELATIONSHIP BETWEEN DUAL SOLUTIONS

| Problem (a) | | Values | P-coefficients | |
|-----|-----|-----|-----|-----|
| Variables | $x_1$ | 20 | 0 | $r_1$ slacks |
|           | $x_2$ | 30 | 0 | $r_2$ |
|           | $u_1$ | 0  | 5 | $y_1$ |
|           | $u_2$ | 40 | 0 | $y_2$ |
| Slacks    | $u_3$ | 0  | 5 | $y_3$ variables |
|           | $u_4$ | 15 | 0 | $y_4$ |
|           | $u_5$ | 20 | 0 | $y_5$ |
|           | P-coefficients | Values | | Problem (b) |

We can put the duality relationship to profitable use by always solving the easier of the two problems. This usually turns out to be the problem which requires no dummy variables to initialise it.

Attempt problems 8.4.

### 8.5 SUMMARY OF CHAPTER 8

If an mxn game cannot be reduced to an rx2 or a 2xs game, then it can be converted into a linear programming problem once its value has been made positive by the addition of a constant to all its elements. In general, the game from B's point of view is the easier to solve since it requires no dummy variables.

To any linear programming problem with a finite solution there exists a dual with the same optimal value, the slacks of one corresponding to the variables of the other.

Attempt problems 8.5.

# Problems

1.1 (a)   If you were a member of a local education authority interested in the building of a new comprehensive school to replace existing local schools, suggest some objectives you might wish to optimise and list some of the constraints that may have to be taken into account by the architects.

   (b)   A farmer has 4 fields he wishes to use for growing root crops and/or barley.   Suggest objectives he might wish to optimise and some possible constraints.

   (c)   Which of the following algebraic expressions and relationships are not linear, and which of these can be made linear?   The (unknown) variables are denoted by z, y, x ....

   (i)    $x - y + 3z$

   (ii)   $x(y + z)$

   (iii)  $x + 2y + 3 \leqslant 6 + 9z + 1$

   (iv)   $x \geqslant 1 + 1/y$

   (v)    $x + u(1 + v) - y + 3z$

   (vi)   $x/y = 1$

   (vii)  $x/y = z$

1.2      On separate graphs plot the regions defined by each of the following:

   (a)   $x \geqslant o, \ y \geqslant o, \ 2x + y \leqslant 3$

   (b)   $x \geqslant 1, \ y \leqslant 5, \ x \leqslant y$

   (c)   $x + y \leqslant 10, \ x + y \geqslant -10, \ x - y \leqslant 10, \ y - x \geqslant 10$

1.3      On the corresponding feasible regions as defined in Problem 1.2, find and mark the following:

   (a)   Maximum of $x + y$ (consider $P = x + y$),

   (b)   Maximum of $2x - 3y$,

   (c)   Minimum of $x + 4y$.

1.4 (a)   By drawing profit lines with gradient $-2/_3$, find the new optimal solution to the problem of the text.

   (b)   Which inequality constraints become equalities at the optimum?

   (c)   By how much do the other constraints differ from equalities at the optimum?

   (d)   Which constraints are *redundant* (i.e. their boundaries lie entirely outside the feasible region)?.

(e) If it were imperative that at least 55 instruments be produced, how would this affect the feasible region? What can you say about solutions to this problem?

(f) Omitting the above constraint, suggest three further 'stock' constraints with the following properties:

(i) Has no effect on the feasible region.

(ii) Changes the feasible region but not the optimal profit.

(iii) Changes both the feasible region and the optimal profit.

1.5 (a) Show that it is the *ratio* of the net profits rather than their separate values which affects the *position* of the optimum.

(b) Assuming the profit on the precision timer remains at £15, what ranges of values for the profit on the standard model will lead to optimal profits at B and at E (Fig.4)?

(c) Solve the following problem:

Maximise $9x + 15y$

subject to $x + y \leqslant 50$

$$2x + y \leqslant 110$$

$$x + 2y \leqslant 30$$

$$x + 5y \leqslant 185$$

$$5x + 6y \geqslant 300$$

$$x \geqslant 0, \ y \geqslant 0$$

1.6 (a) Which of the results discussed in the text remain true in the following circumstances? Give suitable examples.

(i) The polygon is no longer convex.

(ii) The polygon is convex but its sides may be curved.

(iii) The polygon is straight edged and convex, but the profit function is curved.

(b) Graph the following feasible region:

$$x + y \geqslant 5$$

$$x - y \leqslant 10$$

$$x - 2y \leqslant 0$$

$$x \geqslant 0, \ y \geqslant 0$$

and plot the value of the profit line $P = x + 3y$ along the length of the boundary.

(c) Consider the following three-dimensional feasible region consisting of the points within and on a cube with a missing vertex (Fig.Pl), defined by

$$0 \leqslant x \leqslant 10, \ 0 \leqslant y \leqslant 10, \ 0 \leqslant z \leqslant 10, \ x + y + z \leqslant 25$$

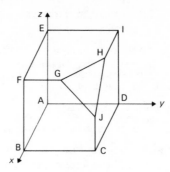

Fig. P. 1

By evaluating the profit at each of its vertices find the maximum of P when

(i)    $P = x + 2y + z$

(ii)   $P = x + y/2 + z$

(iii)  $P = x + y$

1.7 (a)  A manufacturer produces two types of radio, the profit on the one being 10% higher than on the other. How many of each should be produced to maximise his profit if his only restrictions are the stock constraints given in Table P1?

TABLE P1  THE STOCK CONSTRAINTS

| Component | Stock | No. used per radio | |
| | | Type 1 (higher profit) | Type 2 |
| --- | --- | --- | --- |
| a | 160 | 2 | 0 |
| b | 320 | 3 | 2 |
| c | 240 | 1 | 3 |
| d | 360 | 2 | 4 |

(b)  Maximise    $P = x + y$

subject to $5y + x \geqslant 50$

$2y + 3x \geqslant 60$

$y - x \leqslant 0$

(c)  I rent an allotment of 300 square metres and I have to decide how much of the three basic types of vegetable (root crops, brassicas, peas and beans) to plant. By talking to other gardeners I have drawn up the following table, showing the approximate time I must spend on each crop and my expected profit.

| | hours/square metre | profit/square metre |
| --- | --- | --- |
| Root crops | 0.30 | 20p |
| Brassicas | 0.12 | 5p |
| Peas & beans | 0.15 | 10p |

Assuming that I need at least 30 square metres of each type, and cannot afford to spend more than 60 hours per year working at the allotment, write out the linear programming problems to

(i)   Maximise my profit,

(ii)  Maximise my profit if I cost my time at £1 per hour.

By substituting from the equality constraint, reduce these problems to ones in two variables (rootcrops and peas & beans), and hence solve by graphical means.

(d)   For a certain manufacturing process a manufacturer requires 1000 kg of nickel and 1600 kg of tin.   He obtains this by buying ordinary scrap metal and scrap cars.   One tonne of scrap metal yields 100 kg of nickel and 400 kg of tin, whilst one tonne weight of cars yields 100 kg of nickel and the same of tin.   If the scrap metal costs £15 per tonne and scrap cars £5 per tonne, how many tonnes of each should he buy, and what will be his capital outlay?

2.1   Consider again the three-dimensional problem of Fig.Pl.

$$0 \leqslant x \leqslant 10$$
$$0 \leqslant y \leqslant 10$$
$$0 \leqslant z \leqslant 10$$
$$x + y + z \leqslant 25$$

(a)   If the optimum is at G and the initial solution is at A, why won't the simplex method go directly to the optimum along the line AG or along the lines AE and EG?

(b)   Using each of the profit lines of 1.6(c), show all the possible paths the simplex method could take to the optimum, when starting from A.

(c)   Write out the constraints of this problem using the slack variables r, s, t, u.

2.2 (a)   In step (iii) of the text, if one of the maximum values obtained for y had been infinite (caused by dividing by zero) would this have affected our choice of y?

(b)   Consider again the problem of 1.7(a):

Maximise    $P = 11x + 10y$

subject to   $x \qquad \leqslant \ 80$

$$3x + 2y \leqslant 320$$
$$x + 3y \leqslant 240$$
$$x + 2y \leqslant 180$$
$$x, y \ \geqslant 0$$

Add in slack variables r, s, t, u to put the problem into standard form with respect to the initial feasible solution $x = 0$, $y = 0$. Complete the first pass through the steps of the simplex method.

2.3   Complete the second pass of the problem started in Problem 2.2.

2.4   Continue Problem 2.2 by completing the third pass.

2.5 (a)   Complete the remaining passes of Problem 2.2, showing all your steps on the graph of the feasible region.

(b)   Maximise      $2x - 4y + 3z$

     subject to     $x - y + z \leqslant 12$

                   $3x + 4y + 5z \leqslant 30$

                   $2x + 3y - z \leqslant 3$

                   $x \geqslant 0, z \geqslant 0$

                   (y unconstrained)

2.6 (a)   By the algebraic simplex method, find the maximum of $P = 2x + y/2 + z$ subject to the constraints of Problem 2.1.

(b)   Maximise      $5x_1 + 4x_2 + 6x_3$

     subject to     $x_1 - x_2 + x_3 \leqslant 20$

                   $3x_1 + 2x_2 + 4x_3 \leqslant 42$

                   $3x_1 + 2x_2 \leqslant 30$

                   $x_1, x_2, x_3 \leqslant 0$

(c)   Maximise      $-3x + 4y - z$

     subject to           $y + z \leqslant 4$

                   $2x + 3y - z \leqslant 8$

                   $-x + 2y + 4z \leqslant 11$

                   $y \geqslant 0, z \geqslant 0$

                   (x unconstrained).

     (Hint: Don't introduce y first)

3.1   Solve Problem 2.2 by the simplex tabular method.   At each stage compare your working with that of Problems 2.2 and 2.5.

3.2   As for Problem 3.1, but solve Problem 2.5 (b), using checks.

3.3 (a)   Solve Problem 2.6 (b) by the simplex tabular method.

(b)   Solve Problem 2.6 (c) *without* following the hint.

4.1 (a)   Suggest another way of bypassing the difficulty of starting the problem in the text by using a substitution.

(b)   Consider the problem of maximising

       $P = x_1 - 2x_2 + x_3$

     subject to   $x_1 - 2x_2 + 3x_3 \qquad + x_5 = 1$           (i)

                  $x_2 - x_3 + x_4 + 2x_5 = 2$           (ii)

         $x_1 \qquad + x_3 \qquad - x_5 = 3$           (iii)

        $x_1, x_2, x_3, x_4, x_5 \geqslant 0$

     Here the constraints are already in the form of equalities.   By inspection $x_1 = 0$, $x_2 = 4$, $x_3 = 3$, $x_4 = 1$, $x_5 = 0$ is a basic feasible feasible solution.   Put the problem into standard form with respect to this basic feasible solution and hence solve the problem.

(c)   Consider again the above problem, and note that $x_4$ and $x_5$ do not occur in the profit function;   hence we can use the constraints to *define* $x_4$ and $x_5$ in terms of $x_1$, $x_2$ and $x_3$ and to find a relationship between $x_1$, $x_2$ and $x_3$.    For example we have

$$x_5 = 1 - x_1 + 2x_2 - 3x_3 \qquad\qquad \text{(ia)(from (i))}$$

$$x_4 = 2x_1 - 5x_2 + 7x_3 \qquad\qquad \text{(iia)(from (i), (iii))}$$

$$x_1 - x_2 + x_3 = 2 \qquad\qquad \text{(iiia)(from (i), (iii))}$$

If we substitute for $x_1$ into the profit equation we obtain

$$P = 2 - x_2 - x_3$$

which is maximised when $x_2 = x_3 = 0$, giving a profit of 2.

Compare with the solution of (b) and comment.

4.2 (a)  Use a dummy variable to prime Problem 4.1(a) and hence solve it.

(b)  Consider the problem and method of dummy variables of the text. Since the introduction of 3 dummy variables usually requires 3 extra tableaux to reach an optimal solution (each dummy must leave the basis), the following method is usually employed to reduce the amount of working when the constraints are of the form $\geqslant$.    Write out the problem involving *just* the slacks.    Now pick out that constraint with the largest constant on the right-hand side. Subtract each of the remaining constraints in turn from this particular one and use these differences in place of the original constraints.    It will be seen that it is now only the original constraint (that with the largest constant term) that requires a dummy variable.    Use this method to find an initial basic feasible solution for the problem of the text and hence find an optimal solution.

(c)  Use dummy variables to find an easy initial basic feasible solution to Problem 4.1(b) and hence solve it.

4.3  Solve the following problems by the tabular method, indicating where you have a choice of pivot row and which of the basic feasible solutions are degenerate.

(a)  Maximise      $x + y + z$

subject to    $2x - 3y + 2z \leqslant 2$

$-3x + 2y + 2z \leqslant 2$

$2x + 2y - 3z \leqslant 2$

$x, y, z \geqslant 0$

(b)  Maximise    $11y + 10x$

subject to $10y + x \leqslant 100$

$5y + x \leqslant 50$

$2y + x \leqslant 20$

$x \leqslant 10$

$x, y \geqslant 0$

4.4  By the simplex tabular method

(a)  Find the *two* optimal solutions of:

$$\text{Minimise} \quad x - 2y + 3z$$

$$\text{subject to} \quad x - y + z + 2r \qquad\qquad = 10$$
$$y - z \qquad + s \quad = 1$$
$$y \quad + 2r \qquad + t = \quad 8$$
$$x, y, r, s, t \geqslant 0$$

(b)  Minimise    $3x + 2y + z$

$$\text{subject to} \quad x + \;\;\; y + 2z \leqslant 6$$
$$x + 4y \qquad\; \geqslant 12$$
$$2x - \;\;\; y + 3z \geqslant 7$$
$$x \qquad\; - z \geqslant 2$$
$$y + 4z \geqslant 20$$
$$x, y, z \geqslant 0$$

(Hint:  Use the method of 4.2(b) to reduce the number of dummy variables.)

(c)  Maximise    $2x + y - z$

$$\text{subject to} \quad -\;\; y + \;\; z \leqslant 4$$
$$3x \qquad + 4z \leqslant 6$$
$$-x - 2y + 3z \leqslant 8$$
$$x, y, z \geqslant 0$$

4.5  This is a trim-loss problem.  A manufacturer of polythene bags buys rolls of polythene sheeting 2.3 metres wide.  From these he cuts narrower rolls of widths 1.20, 0.80 and 0.45 metres as required for the bags he produces.  On a particular day, he gets orders which would use 30 of the 1.20 metre rolls, 40 of the 0.80 and 50 of the 0.45.  How best should he cut the rolls so as to minimise his trim loss?  For example, he could set the knives so as to get 1 of the 1.20 metre and 2 of the 0.45 metre rolls per 2.3 metre roll, and be left with a roll of width 0.20 metres which he would have to discard as waste.

(a)  Write out the 5 possible knife settings he could use to produce rolls of the required widths, and calculate the waste produced by each.

(b)  Denoting the number of rolls produced at each of these settings by $x_1 \ldots x_5$ respectively, write out the linear programming problem to minimise waste, assuming any over-production can be used on later days.

5.1  Consider a problem similar to the one discussed in the text in which the retailers send in orders of 52, 0, 35 and 41 generators respectively.  Assuming that the travelling costs and stocks remain as in the text, write out the linear programming problem and find an initial feasible solution by using the cheapest routes first. Check whether this solution is basic.  Evaluate the corresponding total cost.

5.2  Using the problem stated in 5.1 evaluate the consequences of bringing the route from depot 1 to retailer 4 $(x_4)$ into the basis.

5.3 Draw the closed circuits associated with each of the zeros of the initial solution to Problem 5.1, using the notation of the text, and hence evaluate the *coefficients* only of the cost function. Deduce the result obtained in Problem 5.2 and suggest which variables should next be brought into the basis.

5.4 Complete the solution of Problem 5.1 and find an alternative solution with the same minimum cost.

5.5 Use Dantzig's method to solve Problem 5.1.

5.6 Suppose that in Problem 5.1, retailer 1 requires only 43 generators, and depot 2 has a stock of 46, the other values remaining as before. Show that the initial solution using the minimum-cost routes is degenerate, and use the '$\varepsilon$' method to find an initial basic feasible solution. Hence solve the problem.

5.7 Considering again Problem 5.1, find the optimal solutions when depot 1 has (a) 50 generators

　　　　　　　　　　(b) 40 generators

while all the other constants remain unchanged.

5.8(a) Table P2 gives the predicted egg production and consumption (in units of 12 000 000) for 12 regions of the U S A in a particular year.

### TABLE P2

| Region | Estimated Production | Estimated Consumption |
|---|---|---|
| 1 | 101 | 121 |
| 2 | 278 | 348 |
| 3 | 51 | 46 |
| 4 | 88 | 106 |
| 5 | 456 | 140 |
| 6 | 928 | 326 |
| 7 | 421 | 452 |
| 8 | 984 | 1036 |
| 9 | 387 | 402 |
| 10 | 236 | 397 |
| 11 | 538 | 690 |
| 12 | 411 | 815 |

(Taken from *The Competitive Position of the Connecticut Poultry Industry* by G.G. Judge, 1956 by kind permission of the publishers, The University of Connecticut.)

Find an optimal routing of eggs from regions with a predicted surplus to regions with a predicted deficit, such that the costs of transportation are minimised. The estimated transportation costs between regions (in cents per dozen eggs) is given in Table P3.

(b) Table P4 shows the cost of processing a given list of jobs at various plants within a large company. The problem is to allocate the jobs (one to each plant) in such a way that the total cost to the company is at a minimum.

### TABLE P3  *TRANSPORTATION COSTS*

| | 1 | 2 | 3 | 4 | 5 | 6 | 7 | 8 | 9 | 10 | 11 | 12 |
|---|---|---|---|---|---|---|---|---|---|---|---|---|
| 1 | 0 | 1.8 | 4.0 | 3.5 | 10.7 | 11.6 | 11.6 | 12.9 | 14.7 | 14.4 | 16.2 | 18.0 |
| 2 | | 0 | 5.1 | 2.4 | 10.0 | 10.1 | 8.1 | 11.3 | 14.5 | 8.4 | 13.0 | 15.2 |
| 3 | | | 0 | 5.1 | 6.7 | 7.7 | 9.4 | 8.4 | 12.1 | 11.7 | 11.6 | 13.5 |
| 4 | | | | 0 | 6.0 | 6.4 | 4.1 | 7.6 | 10.8 | 6.2 | 9.2 | 10.3 |
| 5 | | | | | 0 | 2.6 | 5.0 | 1.7 | 5.0 | 5.3 | 4.4 | 5.6 |
| 6 | | | | | | 0 | 2.4 | 1.2 | 4.4 | 4.1 | 2.9 | 5.1 |
| 7 | | | | | | | 0 | 3.3 | 5.3 | 3.9 | 6.4 | 8.3 |
| 8 | | | | | | | | 0 | 3.5 | 3.9 | 2.7 | 3.9 |
| 9 | | | | | | | | | 0 | 1.8 | 1.7 | 5.3 |
| 10 | | | | | | | | | | 0 | 3.5 | 5.3 |
| 11 | | | | | | | | | | | 0 | 2.4 |
| 12 | | | | | | | | | | | | 0 |

(column header: *Regions*; row header: *Regions*)

### TABLE P4  *JOB COSTS*

Plant

| Job | $P_1$ | $P_2$ | $P_3$ | $P_4$ |
|---|---|---|---|---|
| $J_1$ | 15 | 14 | 14 | 17 |
| $J_2$ | 16 | 17 | 18 | 17 |
| $J_3$ | 12 | 11 | 10 | 9 |
| $J_4$ | 10 | 8 | 11 | 10 |

### TABLE P5  *A POSSIBLE ALLOCATION*

| | $P_1$ | $P_2$ | $P_3$ | $P_4$ |
|---|---|---|---|---|
| $J_1$ | 1 | 0 | 0 | 0 |
| $J_2$ | 0 | 0 | 1 | 0 |
| $J_3$ | 0 | 1 | 0 | 0 |
| $J_4$ | 0 | 0 | 0 | 1 |

A possible allocation might be denoted as in Table P5, with an associated cost of 54 units, where the 1s represent the actual allocations used (one to each row and column) and the 0s those not used.

(i)  If we let $x_{ij}$ take the value 1 if job $J_i$ is allocated to plant $P_j$ and 0 otherwise, show that the conditions

$$x_{11} + x_{12} + x_{13} + x_{14} = 1$$

and $x_{11}, x_{12}, x_{13}, x_{14}$ are non-negative integers
imply that just one of $x_{11}$ to $x_{14}$ is 1 and the others zero.

(ii) Using (i) and the fact that the solution to a transportation problem consists only of non-negative integers, deduce that the allocation problem can be solved by the transportation algorithm.  Hence solve the problem paying particular attention to its inherent degeneracy.

6.1(a)  Which of the following are competitive?  Who are the competitors?

(i)  A game of poker.

(ii)  A motor-car race.

(iii) The 1914-18 war.

(iv)  A game of patience.

(v)  A marriage.

(vi)  Gardening.

(vii) This course.

6.1(b)   State the number of *players* (if possible) in the following
         *games*:

    (i)    A game of cricket.

    (ii)   A horse race.

    (iii) 1914-18 war.

    (iv)  National politics.

6.2(a)   Suggest possible *utilities* for the following games:

    (i)    Noughts and crosses.

    (ii)   Party politics.

    (iii) Patience.

    (iv)  Darts.

    (v)    Strip poker.

  (b)   Which of the following are *zero-sum* games?

    (i)    Chess

    (ii)   Party politics

    (iii) Patience

    (iv)  Cricket

    (v)    Horse-betting.

6.3      Consider the following simple game.   Players A and B independently
         write down one of the numbers 1 or 2.    If the numbers are found to
         be equal A wins 1p from B, otherwise B wins 1p from A.    Suggest
         possible *strategies* that A and B might use.

6.4      With reference to Example 2 (page 35 ):

    (i)    What is the payoff matrix for B?

    (ii)   From this table find $V(B)$, the value of the game to B.

    (iii) Can you suggest what the relationship between $V(A)$ and $V(B)$
         is in a 2 person, zero-sum game with an equilibrium value?
         Can you prove it?

6.5      In a particular 2 person, zero-sum game, the payoff matrix to A is
         given in Table P6.

*TABLE* P6

| B :<br>A | $B_1$ | $B_2$ | $B_3$ |
|---|---|---|---|
| $A_1$: | 1 | 3 | 2 |
| $A_2$: | 0 | 4 | -6 |

Analyse this game:

    (i)    How should A play?

    (ii)   How should B play?

    (iii) What is the value of the game?

      (iv)   Is it fair?

      (v)    Is it in equilibrium?   If so demonstrate it.

6.6     Consider again Example 3 (page 36) with the matrix of Table P7.

TABLE P7

| B : | $B_1$ | $B_2$ | $B_3$ |
|---|---|---|---|
| A | | | |
| $A_1$ : | 2 | 4 | 6 |
| $A_2$ : | -2 | -4 | 6 |
| $A_3$ : | 0 | -2 | -4 |

      (i)    What is the maximin value of the game?

      (ii)   What is the minimax value of the game?

      (iii) What is the saddle point?

      (iv)   What is the value of the game?

      (v)    What is A's optimum strategy?

      (vi)   What is B's optimum strategy?

6.7     Is Example 2 (page 35) a game of full information?

6.8     (i)    Construct a payoff matrix for Problem 6.3

      (ii)   What are the minimax and maximin values of this game?

      (iii) Is this game stable?

      (iv)   Has this game a solution in pure strategies?

6.9     (i)    What are the minimax and maximin values of the game whose payoff matrix is shown in Table P8.

TABLE P8

| B : | $B_1$ | $B_3$ | $B_3$ |
|---|---|---|---|
| A | | | |
| $A_1$ : | -1 | 1 | 2 |
| $A_2$ : | 3 | -4 | 0 |
| $A_3$ : | -2 | -2 | -4 |

      (i)    Reduce this payoff matrix

      (ii)   What are the minimax and maximin values of the reduced game?

      (iii) Has this game a solution in pure strategies?  If so, what is it?

6.10(a) Solve the following game (Table P9)

    (b) Show that the general 2 x 2 game with a saddle point (Table P10) must have a dominance.  This result is not true for higher order games.

| TABLE P9 | | | |
|---|---|---|---|
| B : | B$_1$ | B$_2$ | B$_3$ |
| A | | | |
| A$_1$: | 1 | 2 | 3 |
| A$_2$: | 0 | 3 | -1 |
| A$_3$: | -1 | -2 | 4 |

TABLE P10 *THE GENERAL 2 x 2 MATRIX GAME*

| B : | B$_1$ | B$_2$ |
|---|---|---|
| A | | |
| A$_1$: | a$_{11}$ | a$_{12}$ |
| A$_2$: | a$_{21}$ | a$_{22}$ |

(c) Two women each own one of a pair of silver ear-rings valued at £4 the pair, the silver in each being worth £2. They each decide that they would like the other half of the pair, so they adopt the following procedure. They each make a sealed bid of £0,1,2 or 3 and the one with the higher bid buys the ear-ring from the other for the amount of the bid. If the bids are equal, nothing is exchanged. Formulate this problem in a zero-sum game matrix assuming that the utilities are proportional to the money involved, and hence decide upon a suitable choice of strategies for the players.

7.1 Explain the statement $-V(A) \geq -\beta$ in the last line of the text. Why do the negative signs occur?

7.2 (a) Why are $p_1$, $p_4$, $p_7$ greater than zero and not greater than *or equal* to zero in equations (1).

(b) Write out the corresponding proof in which B adheres to his optimum strategy and A changes from his.

7.3 (a) Why must the value of the game to A lie between -3 and +2?

(b) Obtain the solution of the 2 x 2 zero-sum game shown in Table P11 by the method of this section.

TABLE P11

| B : | B$_1$ | B$_3$ |
|---|---|---|
| A | | |
| A$_1$: | -1 | 1 |
| A$_2$: | 3 | -4 |

(c) Obtain the solution of the general 2 x 2 zero-sum game, using the notation of Table P10, assuming there to be no saddle point.

7.4 (a) Which lines on Fig.11 (page **43**) correspond to the thick lines on Fig.10?

(b) What feature of the payoff matrix for Example 9 (page 44) makes the graphical solutions identical for A and B?

(c) Obtain the solution of Problem 7.3(b) by graphical methods.

7.5 (a) "B will not use more strategies than A". Explain this more fully.

(b) Complete the example in the text (find $q_1$, $q_3$, V).

7.6 (a) Solve the following game (Table P12).

(b)   A has two cards, marked 2 and 4, and B has five cards, marked 1,2,3, 4 and 5.   They each choose a card simultaneously.   If player A's choice is 1 greater than B's he wins £a and if 1 less he loses £(b-a) where b > a > 0.   Otherwise no money changes hands.   Find the optimal strategies for A and B and the value of the game.

*TABLE* P12

| B : | $B_1$ | $B_2$ | $B_3$ |
| --- | --- | --- | --- |
| A | | | |
| $A_1$: | 3 | 0 | 0 |
| $A_2$: | 1 | 2 | 2 |
| $A_3$: | -1 | 4 | 3 |
| $A_4$: | 4 | 1 | 0 |

8.1   How would result (3) of the text be altered if we were to assume that V, the value of the game to A, were negative?

(Put V = = -U, where U>0)

8.2   What is the dual problem if we assume V < 0?

8.3   What is B's optimum strategy?

8.4 (a)   In each of the following, state the dual of the given problem, solve *one* of them, and hence deduce the optimal solution to the other.

(i)   Maximise     $P = x + y + 2z$

subject to         $x + 2y \qquad \leqslant 10$

$x - y + \ z \leqslant 20$

$2x + y + 4z \leqslant 16$

$x, y, z \geqslant 0$

(ii)   Maximise     $P = x_1$

subject to         $x_1 + x_2 + x_3 = 6$

$x_1 \qquad +2x_3 \geqslant 4$

$x_1 - x_2 \qquad \geqslant 0$

$x_1, x_2, x_3 \ \geqslant 0$

(b)   Solve the following problem *and* its dual:

Minimise     $P = x_1 - x_2$

subject to         $x_1 + x_2 \geqslant 5$

$2x_1 - x_2 \geqslant 4$

$3x_1 +2x_2 \leqslant 6$

$x_1, x_2 \geqslant 0$

8.5 (a)   Find the value of the game shown in   Table P13   and the optimal strategies for both A and B.

(b)   X, Y and Z are three towns lying in a straight line such that Y is 5 miles from each of X and Z.   Each of two players A and B can go

to any of the three towns.   Player A wishes to go to the same town
as B, but B wishes to avoid A as far as possible.   If they go to
the same town B pays A £10, otherwise A pays B £1 for each mile they
are apart.

(i)    Set up the payoff matrix.

(ii)   Find the optimal strategies for A and B and the value of the
       game.

(c)  Blue secretly picks up one of the numbers 1, 2 or 3.    Red proceeds
to guess that number.   Each time Red announces his guess, Blue
answers 'high', 'low' or 'correct'.   The game continues until Red
has guessed correctly.   The payoff to Blue is two less than the
number of guesses required by Red to identify the number.   How
should Blue and Red play and what is the outcome?

### TABLE P13

| B:<br>A | $B_1$ | $B_2$ | $B_3$ |
|---|---|---|---|
| $A_1$: | 3 | 4 | -7 |
| $A_2$: | -2 | 1 | 6 |
| $A_3$: | 5 | 5 | -8 |

# Solutions

1.1(a)  Possible objectives : minimise the costs or maximise the facilities subject to an upper cost limit.  Constraints : it must be capable of taking the number of pupils (present and predicted) within its catchment area;  satisfy local fire regulations;  be within easy reach of all the areas it serves;  be able to serve as a local community centre during the evenings, etc., etc.

(b)  Possible objectives : maximise the profit on those 4 fields or maximise the profit on the whole farm (i.e. allowing the crops to be used as food for his own livestock, etc.).  Possible constraints : account may have to be taken of crops previously planted (crop rotation);  the availability of cattle fodder and the estimated cost of buying in more;  the cash flow situation, bearing in mind that seed potatoes, for example, are very expensive and that some crops need more fertilizer than others, etc., etc.

(c)  (i)  Linear

(ii)  Non-linear, since when expanded it is $xy + xz$.

(iii)  A relationship between two linear expressions.

(iv)  Non-linear and remaining so even when rearranged as $xy \geqslant y + 1$.

(N.B. The multiplication of both sides of an inequality by a negative number always reverses the sense of the inequality).

(v)  Non-linear, since, when expanded, it contains $uv$.

(vi)  Non-linear as it stands, but if rearranged becomes $x = y$, ($y \neq 0$) which is linear.

(vii) Non-linear even when rearranged.

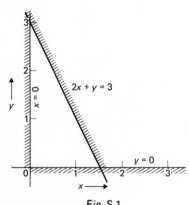

Fig. S.1

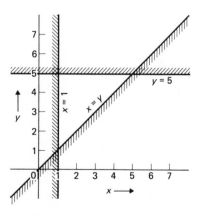

Fig. S.2

1.2 (a)  See Fig.S1.    The origin (O.O) satisfies $2x + y \leqslant 3$, so we reject
points on the opposite side of the boundary $2x + y = 3$.

(b)  See Fig.S2.    The point (O.5) satisfies $x \leqslant y$, therefore the points
on the opposite side of the line $x = y$ are rejected.

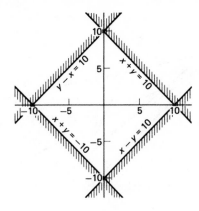

Fig. S.3

(c)  See Fig.S3.    The origin can be used to test all the inequalities.

1.3      See Figs.S4, S5 and S6.

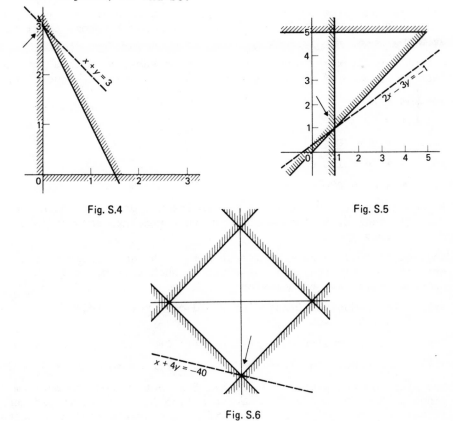

Fig. S.4                              Fig. S.5

Fig. S.6

1.4

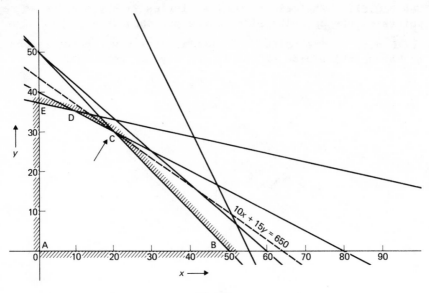

Fig. S.7

(a)  The new optimum is at the point C (Fig.S7) as the profit increases
     with the distance of the line 10x + 15y = P from A.

(b)  C is the intersection of the two constraint boundaries x + y = 50
     and x + 2y = 80, so that constraints (2) and (5) become equalities
     at C.

(c)  C is the point x = 20, y = 30, so we must substitute this into the
     left-hand sides of the other constraints to answer the question.

$$(3) \begin{cases} x = 20 \geqslant 0 \\ y = 30 \geqslant 0 \end{cases}$$      Difference  :   20
                                                                          30

     (4)   2x + y = 70 ⩽ 110                                        40

     (6)    x + 5y = 170 ⩽ 185                                      15

     (7)   5x + 6y = 280 ⩽ 300                                      20

(d)  By inspection the boundary of constraint (7) does not touch the
     feasible region, and hence need not be taken into account in our
     calculations.

(e)  Any solution would have to lie in x + y ⩾ 55 as well as in the
     present feasible region:   there are no such points, hence the
     problem would have no solution.

(f)  (i)    Any redundant constraint would do;   for example,
            4x + 3y ⩽ 400.

     (ii)   Any constraint not 'cutting off' C;   for example, x ⩽ 40.

     (iii)  Any constraint 'cutting off' C;   for example, 5x + 4y ⩽ 200.

1.5(a)  The slope of the profit line is all that determines the optimum,
        since we are concerned with moving a line parallel to itself until
        it has only one point in common with the reasible region.   So that

once the feasible region has been defined, it is only the slope of the profit line (the negative of the ratio of the individual profits) that determines the position of the optimum.    The actual values determine the *numerical* value of the profit.

(b)  Suppose P = Sx + 15y.    At E (Fig.S7) we have the intersection of the lines x = 0, x + 5y = 185, so that any profit line with a gradient between those of the two boundaries will 'end-up' at B. The two gradients are ∞ and -0.2.    However, we can only increase the gradient from -0.2 to 0 and still have non-negative coefficients, hence the range of values for S is from 3 (gradient -0.2) to 0 (gradient 0).

At B, we have the intersection of the lines y = 0 (zero gradient) and x + y = 50 (gradient -1).    We can only decrease the gradient from -1 to -∞ without obtaining negative profits, so that the range of values for S is 15 (gradient -1) to ∞ (gradient -∞).    The latter is, of course, equivalent to P = Cx where C is any positive value.

If you find the preceeding arguments taxing your understanding of geometry and concept of infinity, just use a ruler on Fig.S7 to find the limits, bearing in mind that in order to keep the cost coefficients non-negative the ruler must only vary between the horizontal and vertical in the way shown in Fig.S8.

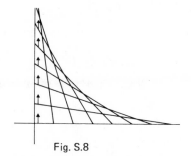

Fig. S.8

(c)  Using the result obtained in the text, and noting that the gradient of the profit function is -9/15 = -0.6, we can solve this problem directly.    For -0.6 lies between -0.5 (the gradient of CD) and -1 (the gradient of CB).    Hence the optimum must still be at C.    The value of the profit is then 9(20) + 15(30) = £630.

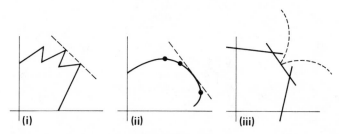

Fig. S.9

1.6(a)   (i)   The optimum will still be a vertex, but the other results no longer hold.  See Fig.S9.

(ii)   The optimum need not be a 'vertex' and so none of the results hold.

(iii) Counter examples can be found to break each of the results. However, by suitably restricting the form of the profit function some of the results can be made to remain valid.

(b)

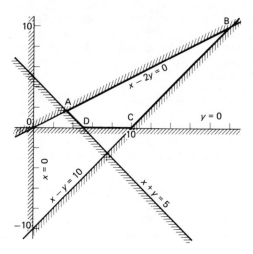

Fig. S.10

The feasible region is the area within ABCD (Fig. S10). The profit at A ($x = 10/3$, $y = 5/3$) is $25/3$, at B ($x = 20$, $y = 10$) is 50, at C ($x = 10$, $y = 0$) is 10, and at D ($x = 5$, $y = 0$) is 5.

Along AB ($x = 2y$) the profit is given by $P = x + 3y = 2y + 3y = 5y$.

Along BC ($x - y = 10$) the profit is given by $P = x + 3y$
$= y + 10 + 3y = 10 + 4y$.

Along CD ($y = 0$) the profit is given by $P = x$.

Along DA ($x + y = 5$) the profit is given by $P = x + 3y = 5 - y + 3y$
$= 5 + 2y$.

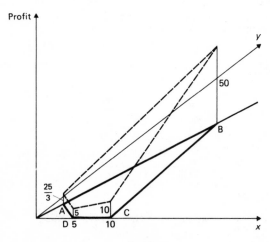

Fig. S.11

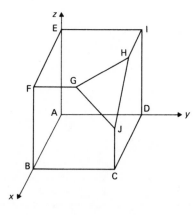

Fig. S.12

Each of these are linear in one variable, and so we can now draw a three-dimensional graph to display the profit on the boundary of the feasible region (Fig.S11).   The feature to note is that the profit function is 'uphill' from D to B along the length of either route.

(c)   With reference to Fig.S12

| Point | x | y | z | $P_{(i)}$ | $P_{(ii)}$ | $P_{(iii)}$ |
|---|---|---|---|---|---|---|
| A | O | O | O | O | O | O |
| B | 10 | O | O | 10 | 10 | 10 |
| C | 10 | 10 | O | 30 | 15 | (20) |
| D | O | 10 | O | 20 | 5 | 10 |
| E | O | O | 10 | 10 | 10 | O |
| F | 10 | O | 10 | 20 | 20 | 10 |
| G | 10 | 5 | 10 | 30 | (22½) | 15 |
| H | 5 | 10 | 10 | (35) | 20 | 15 |
| I | O | 10 | 10 | 30 | 15 | 10 |
| J | 10 | 10 | 5 | (35) | 20 | (20) |

where $P_{(i)}$   $= x + 2y + z$   -   maxima at H and J

$P_{(ii)}$   $= x + \frac{1}{2}y + z$   -   maximum at G.

$P_{(ii)}$   $= x + y$       - maxima at C and J.

1.7(a)   Suppose the profit on the cheaper radio is 1 unit, then the profit on the more expensive is 1.1 units.   If he manufactures y of the first kind and x of the second, then the profit   function is $P = 1y + 1.1x$.

The constraints are

$2x + 0y \leqslant 160$

$3x + 2y \leqslant 320$

$1x + 3y \leqslant 240$

$2x + 4y \leqslant 360$

and, presumably, $x, y \geqslant 0$.

Plotting this feasible region we obtain Fig.S13.   The profit is maximum when passing through D (x = 70, y = 55), the value being 132 units.   Hence the manufacturer should produce 55 of type 1 and 70 of type 2.

(b)   The feasible region is 'open ended' as shown in Fig.S14;   hence the profit line can be moved infinitely far in the direction indicated, giving an infinite profit.   Open ended feasible regions do not usually occur in practice;   they are normally the result of an ill-defined problem.

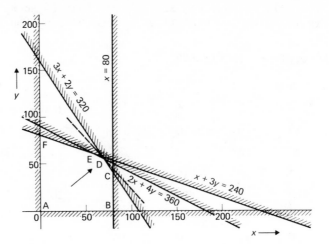

Fig. S.13

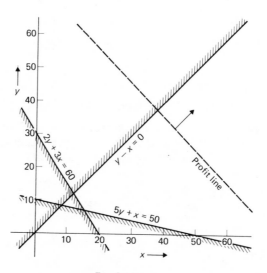

Fig. S.14

(c)   Suppose I plant x, y and z square metres of root crops, brassicas
      and peas & beans respectively, then the constraints are as follows:

$x \geqslant 30, \, y \geqslant 30, \, z \geqslant 30$

$0.30x + 0.12y + 0.15z \leqslant 60$

$x + y + z = 300$

(i)   The profit is $20x + 5y + 10z$ pence.

(ii)  The profit allowing for my time is
      $20x + 5y + 10z - 100 \, (0.3x + 0.12y + 0.15z) = -10x - 7y - 5z.$

This is a problem in three dimensions, but if we use the equality
constraint to eliminate y, we reduce the problem to one in two

dimensions, which can be solved by graphical methods.

We have $y = 300 - x - z$ so that the constraints are

$x \geqslant 30$, $x + z \leqslant 270$, $z \geqslant 30$

$0.18x + 0.03z \leqslant 24$  or  $6x + z \leqslant 800$

The profit functions are

(i)   $15x + 5z + 1500$

(ii)  $-3x + 2z - 2100$

The first is maximised at A, (Fig.S15) which corresponds to
$x = 106$, $z = 164$ (giving $y = 30$) and $P = 3910$ pence.  The second is
maximised at B, corresponding to $x = 30$, $z = 240$ (and hence $y = 30$),
giving a profit of $-1710$ pence.  Anyway, home grown food tastes
better!!

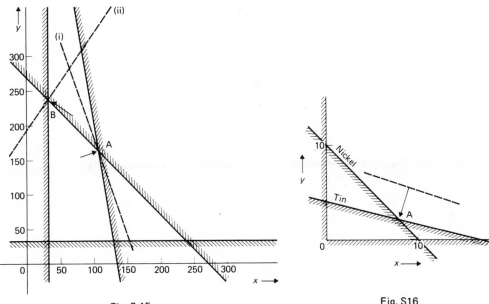

Fig. S.15                    Fig. S16

(d)   Suppose he buys x tonnes of cars and y tonnes of scrap metal, then
we can write the constraints as follows:

nickel : $100x + 100y \geqslant 1000$

tin    : $100x + 400y \geqslant 1600$

        $x, y \geqslant 0$

cost   : $5x + 15y$

The cost function is minimised at A, (Fig.S16) which corresponds to
the solution

$x = 8$ tonnes, $y = 2$ tonnes, cost $= £70$.

2.1(a)  The simplex method searches for the optimum by moving from one
vertex to a neighbouring one along a connecting edge.   Neither AG
nor EG (Fig.P1) are edges.

(b)

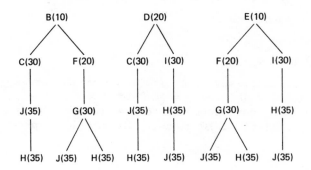

Fig. S.17

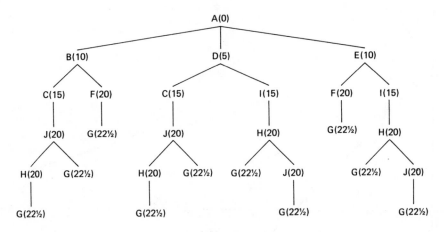

Fig. S.18

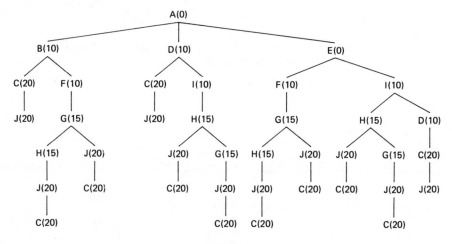

Fig. S.19

In its search for an optimum, the simplex method uses a 'hill climbing' strategy, that is, it never moves to a new vertex which has a lower profit than the one it is already at. $P_{(i)}$ starts at A with profit 0 (A(0)). It can then move to B(10), D(20) or E(10) and so on as displayed in Fig.S17. Similarly for P(ii) starting at A(0) (Fig.S18) and for P(iii) starting at A(0), see Fig.S19.

(c)  $x \geqslant 0, \quad y \geqslant 0, \quad z \geqslant 0$

$x \leqslant 10 : \quad x + r = 10, \qquad\qquad\qquad r \geqslant 0$

$y \leqslant 10 : \quad y + s = 10, \qquad\qquad\qquad s \geqslant 0$

$z \leqslant 10 : \quad z + t = 10, \qquad\qquad\qquad t \geqslant 0$

$x + y + z \leqslant 25 \qquad x + y + z + u = 25, \qquad u \geqslant 0$

2.2(a)  No, since any value chosen would still have satisfied this constraint.

(b)  Note that since we are only concerned with the ratio of the 2 product profits, any multiple of the profit function can be considered (equivalent to changing the monetary unit), hence instead of using $P = 1.1x + y$ as in Problem 1.7, we could have used $P = 11x + 10y$, and divided the final cost by 10.

Problem:

maximise    $P = 11x + 10y$ (a)

subject to    $x \qquad\quad + r \qquad\qquad\qquad = 80$ (b)

$3x + 2y \quad + s \qquad\qquad = 320$ (c)

$x + 3y \qquad\quad + t \qquad = 240$ (d)

$x + 2y \qquad\qquad\quad + u = 180$ (e)

$x, y, r, s, t, u \geqslant 0$

This is in standard form with respect to $(x = 0, y = 0)$, since only these variables occur in the profit function, the other variables each occur in one and only one equation with unit coefficient, and the right-hand sides (constants) are non-negative. For the first pass:

(i)  $(x = 0, y = 0)$ is a vertex of the feasible region (Fig.S20).

(ii)  The value of P at this vertex is zero.

(iii)  The profit equation shows that each unit increase in x produces 11 units increase in P, whereas each unit increase in y produces only 10 units in P. Hence we try to introduce as many x as possible, i.e. we move to the next vertex on $y = 0$. To find the maximum allowable increase in x, set each variable to zero in turn (with $y = 0$)

$r = 0 : \quad x = 80$

$s = 0 : \quad x = 106\frac{2}{3}$

$t = 0 : \quad x = 240$

$u = 0 : \quad x = 180$

This shows that 80 is the largest value for x which satisfies all the constraints.

(iv)   The next vertex is therefore defined by $y = 0$, $r = 0$.

2.3   Since the new vertex is defined by $y = 0$, $r = 0$, all the equations must be written in standard form with respect to this vertex, using equation (b) of Solution 2.2 : $x = 80 - r$.

$$P = 880 \qquad + 10y - 11r \tag{a1}$$

$$x \qquad + \quad r \qquad\qquad = 80 \tag{b1}$$

$$2y - \quad 3r + s \qquad = 80 \tag{c1}$$

$$3y - r \qquad\quad + t \qquad = 160 \tag{d1}$$

$$2y - r \qquad\qquad\quad + u = 100 \tag{e1}$$

Only by increasing y can we increase P from its present value of 880, and so we must determine by how much it can be increased whilst still keeping r zero.

$x = 0$ : no information (equivalent to $\infty$ in Solution 2.2(a)).

$s = 0$ : $y = 40$

$t = 0$ : $y = 53\frac{1}{3}$

$u = 0$ : $y = 50$

Hence we must take $y = 40$, giving $s = 0$, $r = 0$ as the new vertex.

2.4   Putting the equations into standard form with respect to $s = 0$, $r = 0$ requires the substitution of $y = 40 + 3r/2 - s/2$ from equation (c1) of Solution 2.3, giving

$$P = 1280 \qquad + 4r \quad - \quad 5s \tag{a2}$$

$$x \quad + \quad r \qquad\qquad = 80 \tag{b2}$$

$$y - 3r/2 + \quad s/2 \qquad = 40 \tag{c2}$$

$$7r/2 - 3s/2 + t \qquad = 40 \tag{d2}$$

$$2r \quad - \quad s \qquad + u = 20 \tag{e2}$$

Arguing as before, we see that we must introduce r, with possible maxima of: 80, (-80/3), 80/7, 10 respectively.   We discount the negative value (inspection of equation (c2) shows why) and conclude that $r = 10$, $u = 0$ with $s = 0$ defines the next vertex.

2.5(a)   The substitution of $r = 10 + s/2 - u/2$ from equation (e2) of Solution 2.4 will put the equations into standard form with respect to $u = 0$, $s = 0$.

$$P = 1320 \qquad\qquad - 3s \qquad - 2u \tag{a3}$$

$$x \qquad + \quad s/2 \quad - \quad u/2 = 70 \tag{b3}$$

$$y \qquad - \quad s/4 \quad + 3u/4 = 55 \tag{c3}$$

$$s/4 + t - 7u/4 = 5 \tag{d3}$$

$$r - \quad s/2 \quad + \quad u/2 = 10 \tag{e3}$$

The equation (a3) shows that no improvement can be made on the present profit of 1320 units.

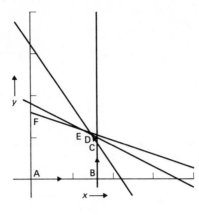

Fig. S.20

(b)  Inserting slacks and putting $y = y_1 - y_2$ (since y is not restricted) we obtain

$$P = 2x - 4y_1 + 4y_2 + 3z$$

subject to
$$x - y_1 + y_2 + z + r = 12$$
$$3x + 4y_1 - 4y_2 + 5z + s = 30$$
$$2x + 3y_1 - 3y_2 - z + t = 3$$
$$x, y_1, y_2, z, r, s, t \geqslant 0$$

The equations are in standard form with respect to $x = 0$, $y_1 = 0$, $y_2 = 0$, $z = 0$. A unit increase in $y_2$ produces the greatest increase in P, hence we must discover by how much $y_2$ may be increased.

Put
$r = 0$:  $y_2 = 12$

$s = 0$:  $y_2 = -7\frac{1}{2}$  (discard)

$t = 0$:  $y_2 = -1$  (discard)

Hence the next vertex is given by $x$, $y_1$, $z$ and $r$ being zero. We must now put the equations into standard form using the first constraint:

$$y_2 = 12 - x + y_1 - z - r.$$
$$P = 48 - 2x - z - 4r$$

subject to
$$x - y_1 + y_2 + z + r = 12$$
$$7x + 9z + 4r + s = 78$$
$$5x - 2z + 3r + t = 39$$

This shows that the optimum has already been obtained, since none of the coefficients in the P equation are positive.

The solution is  $y_2 = 12$, $s = 78$, $t = 39$, $x = 0$, $y_2 = 0$, $z = 0$, $r = 0$, giving $x = 0$, $y = -12$, $z = 0$ in terms of the original variables.

2.6(a)  Maximise    $P = 2x + \frac{1}{2}y + z$

subject to  $x \qquad\qquad + r \qquad\qquad\qquad = 10$

$\qquad\qquad\quad y \qquad\qquad + s \qquad\qquad = 10$

$\qquad\qquad\qquad z \qquad\qquad + t \quad = 10$

$\qquad\quad x + \ y + z \qquad\qquad\qquad + u = 25$

$\qquad\quad x,\ y,\ z,\ r,\ s,\ t,\ u \geqslant 0$

This is in standard form with respect to $x = y = z = 0$.   We should introduce as many x's as possible :

Put    $r = 0$ :  $x = 10$

$\qquad s = 0$ :  no information

$\qquad t = 0$ :  no information

$\qquad u = 0$ :  $x = 25$

We must therefore take x to be 10, and our new vertex is defined by $y = z = r = 0$.   The new standard form is obtained by substituting $x = 10 - r$, giving

maximise    $P = 20 \qquad + \frac{1}{2}y + z - 2r$

subject to  $\qquad\qquad x \qquad + r \qquad\qquad\qquad = 10$

$\qquad\qquad\qquad y \qquad\qquad + s \qquad\qquad = 10$

$\qquad\qquad\qquad\quad z \qquad\qquad + t \qquad = 10$

$\qquad\qquad\quad y + z - r \qquad\qquad + u = 15$

Now introduce as many z's as possible.

Put    $x = 0$ :  no information

$\qquad s = 0$ :  no information

$\qquad t = 0$ :  $z = 10$

$\qquad u = 0$ :  $z = 15$,

leaving $z = 10$, $t = 0$.   The new standard form is

maximise    $P = 30 \qquad + \frac{1}{2}y \qquad - 2r \qquad - t$

subject to  $\qquad\qquad x \qquad + r \qquad\qquad\qquad = 10$

$\qquad\qquad\qquad y \qquad\qquad + s \qquad\qquad = 10$

$\qquad\qquad\qquad\quad z \qquad\qquad + t \qquad = 10$

$\qquad\qquad\qquad y \qquad - r \qquad - t + u = \ 5$

We can now only introduce y's.

Put    $x = 0$ :  no information

$\qquad s = 0$ :  $y = 10$

$\qquad z = 0$ :  no information

$\qquad u = 0$ :  $y = 5$,

so we must take $y = 5$.

The new vertex is defined by $r = 0$, $t = 0$, $u = 0$ and the standard form for the equations becomes

$$\text{maximise} \quad P = 32\tfrac{1}{2} \qquad\qquad -3r/2 \quad -t/2 - u/2$$

$$
\begin{array}{llll}
\text{subject to} & x & +\ r & = 10 \\
& & r + s + t \quad - u & = 5 \\
& z & +\ t & = 10 \\
& y \qquad - r \quad - t \quad + u & = 5
\end{array}
$$

This shows the optimal solution has been attained:

$x = 10$, $y = 5$, $z = 0$, $P = 32\tfrac{1}{2}$

(b)  The problem, incorporating slacks, and in standard form with respect to the initial feasible solution $x_1 = x_2 = x_3 = 0$ is

$$\text{maximise} \quad P = 5x_1 + 4x_2 + 6x_3$$

$$
\begin{array}{l}
\text{subject to} \quad x_1 - x_2 + x_3 + r \qquad\qquad = 20 \\
\qquad\qquad\ \ 3x_1 + 2x_2 + 4x_3 \qquad + s \qquad = 42 \\
\qquad\qquad\ \ 3x_1 + 2x_2 \qquad\qquad\qquad\quad + t = 30 \\
\qquad\qquad\ \ x_1,\ x_2,\ x_3,\ r,\ s,\ t \geqslant 0
\end{array}
$$

We introduce $x_3$ first:

$$
\begin{array}{lll}
\text{Put} & r = 0 : & x_3 = 20 \\
& s = 0 : & x_3 = 10\tfrac{1}{2} \\
& t = 0 : & \text{(no information)}
\end{array}
$$

so that $x_1 = x_2 = s = 0$ in the new vertex.

Substituting for $x_3$ in terms of $x_1$, $x_2$, $s$ from the second constraint gives the equations in standard form:

$$\text{maximise} \quad P = 63 + \tfrac{1}{2}x_1 \quad + \ x_2 \qquad\qquad -3s/2$$

$$
\begin{array}{l}
\text{subject to} \quad x_1/4 - 3x_2/2 \qquad + r - \quad s/4 \qquad = 19/2 \\
\qquad\qquad\ \ 3x_1/4 + \ x_2/2 + x_3 \qquad + \quad s/4 \qquad = 21/2 \\
\qquad\qquad\ \ 3x_1 \quad + 2x_2 \qquad\qquad\qquad + t = 30
\end{array}
$$

The profit equation shows that the profit can be increased by introducing either $x_1$ or $x_2$, but since the latter has the larger coefficient, this is the one that would normally be chosen.  How many?   Putting the non-zero variables zero one at a time gives $x_2 = -19/3$, 21, and 15 respectively, the latter being the one which we must choose.   The next vertex is therefore defined by $x_1 = s = t = 0$.   Putting the equations into standard form with the aid of the last equation gives

$$\text{maximise} \quad P = 78 - \ x_1 \qquad\qquad\qquad -3s/2 - \ t/2$$

$$
\begin{array}{l}
\text{subject to} \quad 5x_1/2 \qquad\qquad + r - \ s/4 + 3t/4 = 32 \\
\qquad\qquad\qquad\qquad x_3 \quad + \ s/4 - \ t/4 = \ 3 \\
\qquad\qquad\ \ 3x_1/2 + x_2 \qquad\qquad\quad + \ t/2 = 15
\end{array}
$$

Since none of the coefficients in the $P$ equation are positive, we have reached an optimal solution:

$x_1 = 0$, $x_2 = 15$, $x_3 = 3$, and $P = 78$.

(c)  Since x is unconstrained, we write $x = x_1 - x_2$ and continue as usual.

Maximise $\qquad$ $P = -3x_1 + 3x_2 + 4y - z$

subject to

$$y + z + r = 4$$
$$2x_1 - 2x_2 + 3y - z + s = 8$$
$$-x_1 + x_2 + 2y + 4z + t = 11$$

These are in standard form with respect to $x_1 = x_2 = y = z = 0$. We would normally insert as many y's as possible for the next vertex, but since we have been warned against this, there is only one other variable, $x_2$, that can be introduced. Putting variables zero one at a time gives $x_2$ = (no information), -4, and 11 respectively. Hence we choose x = 11. This corresponds to the vertex $x_1 = y = z = t = 0$. The standard form for the equations then becomes

maximise $\qquad$ $P = 33 - 2y - 13z - 3t$

subject to

$$y + z + r = 4$$
$$7y + 7z + s + 2t = 30$$
$$-x_1 + x_2 + 2y + 4z + t = 11$$

which shows that the solution has been attained: x = -11, y = 0, z = 0 and P = 33. If we had first introduced y as usual, we should have needed 4 passes through the simplex method, instead of the 1 needed here. Hence it doesn't always pay to take the largest coefficient to determine the variable to be introduced. However, more often than not it has been found to be a good indicator of a quick solution.

3.1

### TABLE S1

| | | x | y | r | s | t | u | | | NOTES |
|---|---|---|---|---|---|---|---|---|---|---|
| 1 | | | | | | | | | | (The numbers in brackets refer to the rows). |
| 2 | P | -11 | -10 | 0 | 0 | 0 | 0 | 0 | | Pivot column with largest negative entry. |
| 3 | r | 1 | 0 | 1 | 0 | 0 | 0 | 80 | 80 | Pivot row with smallest positive entry. |
| 4 | s | 3 | 2 | 0 | 1 | 0 | 0 | 320 | $320/_3$ | |
| 5 | t | 1 | 3 | 0 | 0 | 1 | 0 | 240 | 240 | Introduce x for r. Pivot already 1. |
| 6 | u | 1 | 2 | 0 | 0 | 0 | 1 | 180 | 180 | |
| 7 | P | 0 | -10 | 11 | 0 | 0 | 0 | 880 | | (7) = (2) + 11.(8) |
| 8 | x | 1 | 0 | 1 | 0 | 0 | 0 | 80 | ∞ | (8) = (3)    '∞' is equivalent to no information. |
| 9 | s | 0 | 2 | -3 | 1 | 0 | 0 | 80 | 40 | (9) = (4) - 3.(8)    y replaces s. |
| 10 | t | 0 | 3 | -1 | 0 | 1 | 0 | 160 | $53\frac{1}{3}$ | (10) = (5) - 1.(8) |
| 11 | u | 0 | 2 | -1 | 0 | 0 | 1 | 100 | 50 | (11) = (6) - 1.(8) |
| 12 | P | 0 | 0 | -4 | 5 | 0 | 0 | 1280 | | (12) = (7) + 10.(14) |
| 13 | x | 1 | 0 | 1 | 0 | 0 | 0 | 80 | 8 | (13) = (8), since y entry already zero. |
| 14 | y | 0 | 1 | $-^3/_2$ | $\frac{1}{2}$ | 0 | 0 | 40 | NEG | (14) = (9)/2 |
| 15 | t | 0 | 0 | $^7/_2$ | $-^3/_2$ | 1 | 0 | 40 | $^{80}/_7$ | (15) = (10) - 3.(14) |
| 16 | u | 0 | 0 | 2 | -1 | 0 | 1 | 20 | 10 | (16) = (11) - 2(14) : r replaces u. |
| 17 | P | 0 | 0 | 0 | 3 | 0 | 2 | 1320 | | (17) = (12) + 4.(21): No negative entries. |
| 18 | x | 1 | 0 | 0 | $\frac{1}{2}$ | 0 | $-\frac{1}{2}$ | 70 | | (18) = (13) - 1.(21) |
| 19 | y | 0 | 1 | 0 | $-\frac{1}{4}$ | 0 | $\frac{3}{4}$ | 55 | | (19) = (14) - $^3/_2$.(21) |
| 20 | t | 0 | 0 | 0 | $\frac{1}{4}$ | 1 | $-^7/_4$ | 5 | | (20) = (15) - $^7/_2$.(21) |
| 21 | r | 0 | 0 | 1 | $-\frac{1}{2}$ | 0 | $\frac{1}{2}$ | 10 | | (21) = (16)/2 |

Following the Table S1, the optimal solution is $P = 1320$, $x = 0$, $y = 55$, the above working corresponding exactly to the passes of Problems 2.2 to 2.5.

3.2 With the notation of Solution 2.5(b), the tabular solution is as in Table S2.

### TABLE S2

| 1 | | x | $y_1$ | $y_2$ | z | r | s | t | | chk | NOTES |
|---|---|---|---|---|---|---|---|---|---|---|---|
| 2 | P | -2 | 4 | (-4) | 3 | 0 | 0 | 0 | 0 | 1 | |
| 3 | r | 1 | -1 | [1] | 1 | 1 | 0 | 0 | 12 (12) | 15 | |
| 4 | s | 3 | 4 | -4 | 5 | 0 | 1 | 0 | 30 NEG | 39 | |
| 5 | t | 2 | 3 | -3 | -1 | 0 | 0 | 1 | 3 NEG | 5 | |
| 6 | P | 2 | 0 | 0 | 7 | 4 | 0 | 0 | 48 | 61 | (6) = (2) + 4.(7) |
| 7 | $y_2$ | 1 | -1 | 1 | 1 | 1 | 0 | 0 | 12 | 15 | (7) = (3) |
| 8 | s | 7 | 0 | 0 | 9 | 4 | 1 | 0 | 78 | 99 | (8) = (4) + 4.(7) |
| 9 | t | 5 | 0 | 0 | 2 | 3 | 0 | 1 | 39 | 50 | (9) = (5) + 3.(7) |

The final solution is therefore at $x = z = 0$, $y = 12$, giving a profit of 48.

### TABLE S3

| | $x_1$ | $x_2$ | $x_3$ | r | s | t | | chk |
|---|---|---|---|---|---|---|---|---|
| P | -5 | -4 | (-6) | 0 | 0 | 0 | 0 | -15 |
| r | 1 | -1 | 1 | 1 | 0 | 0 | 20 20 | 22 |
| s | 3 | 2 | [4] | 0 | 1 | 0 | 42 (10½) | 52 |
| t | 3 | 2 | 0 | 0 | 0 | 1 | 30 ∞ | 36 |
| P | -½ | (-1) | 0 | 0 | 3/2 | 0 | 63 | 63 |
| r | ¼ | -3/2 | 0 | 1 | -¼ | 0 | 19/2 NEG | 9 |
| $x_3$ | ¾ | ½ | 1 | 0 | ¼ | 0 | 21/2 21 | 13 |
| t | 3 | [2] | 0 | 0 | 0 | 1 | 30 (15) | 36 |
| P | 1 | 0 | 0 | 0 | 3/2 | ½ | 78 | 81 |
| r | 5/2 | 0 | 0 | 1 | -¼ | ¾ | 32 | 36 |
| $x_3$ | 0 | 0 | 1 | 0 | ¼ | -¼ | 3 | 4 |
| $x_2$ | 3/2 | 1 | 0 | 0 | 0 | ½ | 15 | 18 |

3.3(a) With the notation of Solution 2.6(b), the solution is given in Table S3. The final solution is at $x_1 = 0$, $x_2 = 15$, $x_3 = 3$, giving a profit of 78.

(b) Using the notation of Solution 2.6(c), but disregarding the hint, we obtain the solution shown in Table S4. This gives the maximum value of P as 33, and shows that it occurs at $x = -11$, $y = 0$, $z = 0$.

| | $x_1$ | $x_2$ | y | z | r | s | t | | | chk |
|---|---|---|---|---|---|---|---|---|---|---|
| P | 3 | -3 | (-4) | 1 | 0 | 0 | 0 | 0 | | -3 |
| r | 0 | 0 | 1 | 1 | 1 | 0 | 0 | 4 | 4 | 7 |
| s | 2 | -2 | [3] | -1 | 0 | 1 | 0 | 8 | (8/3) | 11 |
| t | -1 | 1 | 2 | 4 | 0 | 0 | 1 | 11 | $^{11}/_2$ | 18 |
| P | $^{17}/_3$ | (-$^{17}/_3$) | 0 | $-^1/_3$ | 0 | $^4/_3$ | 0 | $^{32}/_3$ | | $^{35}/_3$ |
| r | $-^2/_3$ | [$^2/_3$] | 0 | $^4/_3$ | 1 | $-^1/_3$ | 0 | $^4/_3$ | (2) | $^{10}/_3$ |
| y | $^2/_3$ | $-^2/_3$ | 1 | $-^1/_3$ | 0 | $^1/_3$ | 0 | $^8/_3$ | NEG | $^{11}/_3$ |
| t | $-^7/_3$ | $^7/_3$ | 0 | $^{14}/_3$ | 0 | $-^2/_2$ | 1 | $^{17}/_3$ | $^{17}/_7$ | $^{32}/_3$ |
| P | 0 | 0 | 0 | 11 | $^{17}/_2$ | (-$^3/_2$) | 0 | 22 | | 40 |
| $x_2$ | -1 | 1 | 0 | 2 | $^3/_2$ | $-^1_2$ | 0 | 2 | NEG | 5 |
| y | 0 | 0 | 1 | 1 | 1 | 0 | 0 | 4 | ∞ | 7 |
| t | 0 | 0 | 0 | 0 | $-^7/_2$ | [$^1_2$] | 1 | 1 | (2) | -1 |
| P | 0 | 0 | 0 | 11 | (-2) | 0 | 3 | 25 | | 37 |
| $x_2$ | -1 | 1 | 0 | 2 | -2 | 0 | 1 | 3 | NEG | 4 |
| y | 0 | 0 | 1 | 1 | [1] | 0 | 0 | 4 | (4) | 7 |
| s | 0 | 0 | 0 | 0 | -7 | 1 | 2 | 2 | NEG | -2 |
| P | 0 | 0 | 2 | 13 | 0 | 0 | 3 | 33 | | 51 |
| $x_2$ | -1 | 1 | 2 | 4 | 0 | 0 | 1 | 11 | | 18 |
| r | 0 | 0 | 1 | 1 | 1 | 0 | 0 | 4 | | 7 |
| s | 0 | 0 | 7 | 7 | 0 | 1 | 2 | 30 | | 47 |

*TABLE* S4

4.1(a)  We could replace x by $x_1$ throughout the problem, where $x_1 = x - 5$,
so that $x \geqslant 5$ is the same as $x_1 \geqslant 0$.   The initial tableau would
then be as in Table S5.

| | $x_1$ | y | r | s | t | u | v | | chk |
|---|---|---|---|---|---|---|---|---|---|
| P | -10 | -15 | 0 | 0 | 0 | 0 | 0 | 50 | 25 |
| r | 1 | 1 | 1 | 0 | 0 | 0 | 0 | 45 | 48 |
| s | 2 | 1 | 0 | 1 | 0 | 0 | 0 | 100 | 104 |
| t | 1 | 2 | 0 | 0 | 1 | 0 | 0 | 75 | 79 |
| u | 1 | 5 | 0 | 0 | 0 | 1 | 0 | 180 | 187 |
| v | 5 | 6 | 0 | 0 | 0 | 0 | 1 | 275 | 287 |

*TABLE* S5

Note the resemblance to the second tableau in the text (Table 7).

(b)  Due to the positioning of the zeros in the constraints, there is (in
this problem) only one possible ordering of the basis shown to the
left of the first tableau (Table S6).   The first three tableau are
used to make the equations into standard form with respect to the
given vertex.

*TABLE* S6

| | $x_1$ | $x_2$ | $x_3$ | $x_4$ | $x_5$ | | | chk | |
|---|---|---|---|---|---|---|---|---|---|
| P | -1 | 2 | -1 | 0 | 0 | 0 | | 0 | |
| $x_2$ | 1 | -2 | 3 | 0 | 1 | 1 | | 4 | Arbitrarily start with $x_3$. |
| $x_4$ | 0 | 1 | -1 | 1 | 2 | 2 | | 5 | |
| $x_3$ | 1 | 0 | [1] | 0 | -1 | 3 | | 4 | $x_3$ element to be non-zero only in this row. |
| P | 0 | 2 | 0 | 0 | -1 | 3 | | | |
| $x_2$ | -2 | [-2] | 0 | 0 | 4 | -8 | | -8 | No need to worry about -8 at this intermediate stage. |
| $x_4$ | 1 | 1 | 0 | 1 | 1 | 5 | | 9 | $x_4$ correct : make $x_2$ pivot. |
| $x_3$ | 1 | 0 | 1 | 0 | -1 | 3 | | 4 | |
| P | (-2) | 0 | 0 | 0 | 3 | -5 | | -4 | |
| $x_2$ | 1 | 1 | 0 | 0 | -2 | 4 | 4 | 4 | The equations are now in standard form with respect to the given initial solution. |
| $x_4$ | 0 | 0 | 0 | 1 | 3 | 1 | ∞ | 5 | |
| $x_3$ | [1] | 0 | 1 | 0 | -1 | 3 | (3) | 4 | |
| P | 0 | 0 | 2 | 0 | 1 | 1 | | 4 | |
| $x_2$ | 0 | 1 | -1 | 0 | -1 | 1 | | 0 | |
| $x_4$ | 0 | 0 | 0 | 1 | 3 | 1 | | 5 | |
| $x_1$ | 1 | 0 | 1 | 0 | -1 | 3 | | 4 | |

The solution is $x_1 = 3$, $x_2 = 1$, $x_3 = 0$, $x_4 = 1$, $x_5 = 0$, giving P = 1.

(c) The profit of 2 obtained in the question is greater than that obtained by the formal tabular method in Solution 4.1(b). This is because the constraints $x_4$, $x_5 \geqslant 0$ have been forgotten. By all means use equations (ia) and (iia) to define $x_4$ and $x_5$, but then $x_4 \geqslant 0$ must be replaced by $2x_1 - 5x_2 + 7x_3 \geqslant 0$ and $x_5 \geqslant 0$ by $1 - x_1 + 2x_2 - 3x_3 \geqslant 0$. So the problem is really one of maximising $P = 2 - x_2 - x_3$,

subject to $\quad x_1 - x_2 + 2x_3 = 2$

$$2x_1 - 5x_2 + 7x_3 \geqslant 0$$

$$x_1 - 2x_2 + 3x_3 \leqslant 1$$

$$x_1, x_2, x_3 \geqslant 0$$

4.2(a) Adding a dummy variable $w_1$ to the last constraint allows us to take $x = y = w = 0$ as our initial vertex. We have to add $Mw_1$ to the profit line so as to ensure that $w_1$ does not appear in the final solution. In order that the equations be in standard form with respect to this vertex, we must subtract M times the last constraint from the new P-equation (Table S7). The solution continues as usual.

TABLE S7

| | x | y | r | s | t | u | v | w | w$_1$ | | | check |
|---|---|---|---|---|---|---|---|---|---|---|---|---|
| | (-10) | | | | | | | | | | | |
| P | (-M) | -15 | 0 | 0 | 0 | 0 | 0 | M | 0 | -5M | | -25 -5M |
| r | 1 | 1 | 1 | 0 | 0 | 0 | 0 | 0 | 0 | 50 | 50 | 53 |
| s | 2 | 1 | 0 | 1 | 0 | 0 | 0 | 0 | 0 | 110 | 55 | 114 |
| t | 1 | 2 | 0 | 0 | 1 | 0 | 0 | 0 | 0 | 80 | 80 | 84 |
| u | 1 | 5 | 0 | 0 | 0 | 1 | 0 | 0 | 0 | 185 | 185 | 192 |
| v | 5 | 6 | 0 | 0 | 0 | 0 | 1 | 0 | 0 | 300 | 60 | 312 |
| w$_1$ | [1] | 0 | 0 | 0 | 0 | 0 | 0 | -1 | 1 | 5 | (5) | 6 |
| P | 0 | (-15) | 0 | 0 | 0 | 0 | 0 | -10 | 10 +M | 50 | | 35 + M |
| r | 0 | 1 | 1 | 0 | 0 | 0 | 0 | 1 | -1 | 45 | 45 | 47 |
| s | 0 | 1 | 0 | 1 | 0 | 0 | 0 | 2 | -2 | 100 | 100 | 102 |
| t | 0 | 2 | 0 | 0 | 1 | 0 | 0 | 1 | -1 | 75 | 37½ | 78 |
| u | 0 | [5] | 0 | 0 | 0 | 1 | 0 | 1 | -1 | 180 | (36) | 186 |
| v | 0 | 6 | 0 | 0 | 0 | 0 | 1 | 5 | -5 | 275 | 45⅚ | 282 |
| x | 1 | 0 | 0 | 0 | 0 | 0 | 0 | -1 | 1 | 5 | ∞ | 6 |

(b)  The text problem with slacks is

maximise $\quad$ P = -4x - 3y

subject to
$$2x + y - u = 50 \qquad \text{(i)}$$
$$x + 2y - v = 40 \qquad \text{(ii)}$$
$$5x + 4y - w = 170 \qquad \text{(iii)}$$

Equation (iii) has the largest constant term, so we subtract (i) and (ii) in turn from it to form two new equations (ia), (iia).

The constraints are now
$$3x + 3y + u - w = 120 \qquad \text{(ia)}$$
$$4x + 2y + v - w = 130 \qquad \text{(iia)}$$
$$5x + 4y - w = 170 \qquad \text{(iii)}$$

The first two constraints are in standard form with respect to u = 0, v = 0, but the third must be augmented with a dummy variable z, say, to find an obvious initial basic feasible solution.  The problem now becomes:

maximise $\quad$ P = 4x - 3y $\qquad$ - Mz

subject to
$$3x + 3y + u - w = 120$$
$$4x + 2y + v - w = 130$$
$$5x + 4y - w + z = 170$$

The initial feasible solution is now u = v = z = 0, and in order to put the equations in standard form we must eliminate the z from the profit equation by using the last constraint, to give

maximise $\qquad$ $P = -x(4 - 5M) - y(3 - 4M) - Mw - 170M$

The complete working is shown in Table S8.

## TABLE S8

|   | x | y | u | v | w | z |   |   | Check |
|---|---|---|---|---|---|---|---|---|---|
| P | (4-5M) | 3-4M | O | O | M | O | -170M | | 7-178M |
| u | 3 | 3 | 1 | O | -1 | O | 120 | 40 | 126 |
| v | [4] | 2 | O | 1 | -1 | O | 130 | (32½) | 136 |
| w | 5 | 4 | O | O | -1 | 1 | 170 | 34 | 179 |
| P | O | (1-3M/2) | O | -1+5M/4 | 1-M/4 | O | -130 -15M/2 | | -129 -8M |
| u | O | $^3/_2$ | 1 | $-\frac{3}{4}$ | $-\frac{1}{4}$ | O | 22½ | 15 | 24 |
| x | 1 | $\frac{1}{2}$ | O | $\frac{1}{4}$ | $-\frac{1}{4}$ | O | 32½ | 65 | 34 |
| w | O | [$^3/_2$] | O | $-^5/_4$ | $\frac{1}{4}$ | 1 | 7½ | (5) | 9 |
| P | O | O | O | (-$^1/_6$) | $^5/_6$ | $-^2/_3+M$ | -135 | | -135 +M |
| u | O | O | 1 | [$\frac{1}{2}$] | $-\frac{1}{2}$ | -1 | 15 | (30) | 15 |
| x | 1 | O | O | $^2/_3$ | $-^1/_3$ | $-^1/_3$ | 30 | 45 | 31 |
| y | O | 1 | O | $-^5/_6$ | $^1/_6$ | $^2/_3$ | 5 | NEG | 6 |
| P | O | O | $^1/_3$ | O | $^2/_3$ | -1+M | -130 | | -130+M |
| v | O | O | 2 | 1 | -1 | -2 | 30 | | 30 |
| x | 1 | O | $-^4/_3$ | O | $^1/_3$ | 1 | 10 | | 11 |
| y | O | 1 | $^5/_3$ | O | $-^2/_3$ | -1 | 30 | | 31 |

(N.B. $-1 + M > 0$)  $\quad$ Unnecessary working is shown within the dotted boundary.

Hence the solution $x = 10$, $y = 30$ was obtained in 4 tableaux compared with the 5 needed in the text.

(c) $\quad$ If we add one of the dummy variables $r$, $s$, $t$, to each of the three equations, we shall be able to start with the obvious initial solution $r = 1$, $s = 2$, $t = 3$ (since the right-hand sides are non-negative). However, in order to eliminate these variables from the final solution we must write the profit function as $P = x_1 - 2x_2 + x_3 - M(r + s + t)$, where M is very large. The problem is set out in Table S9; the first step of subtracting M times each constraint from the profit line has already been taken.

TABLE S9

| | $x_1$ | $x_2$ | $x_3$ | $x_4$ | $x_5$ | r | s | t | | | Check |
|---|---|---|---|---|---|---|---|---|---|---|---|
| P | $-1-2M$ | $2+M$ | $\boxed{-1-3M}$ | $-M$ | $-2M$ | O | O | O | $-6M$ | | $O-13M$ |
| r | 1 | $-2$ | $\boxed{3}$ | O | 1 | 1 | O | O | 1 | ⟨$1/3$⟩ | 5 |
| s | O | 1 | $-1$ | 1 | 2 | O | 1 | O | 2 | NEG | 6 |
| t | 1 | O | 1 | O | $-1$ | O | O | 1 | 3 | 3 | 5 |
| P | $\boxed{-2/3-M}$ | $4/3-M$ | O | $-M$ | $1/3-M$ | $1/3+M$ | O | O | $1/3-5M$ | | $5/3-8M$ |
| $x_3$ | $\boxed{1/3}$ | $-2/3$ | 1 | O | $1/3$ | $1/3$ | O | O | $1/3$ | ⟨1⟩ | $5/3$ |
| s | $1/3$ | $1/3$ | O | 1 | $7/3$ | $1/3$ | 1 | O | $7/3$ | 7 | $23/3$ |
| t | $2/3$ | $2/3$ | O | O | $-4/3$ | $-1/3$ | O | 1 | $8/3$ | 4 | $10/3$ |
| P | O | $\boxed{-3M}$ | $2+3M$ | $-M$ | 1 | $1+2M$ | O | O | $1-4M$ | | $5-3M$ |
| $x_1$ | 1 | $-2$ | 3 | O | 1 | 1 | O | O | 1 | NEG | 5 |
| s | O | 1 | $-1$ | 1 | 2 | O | 1 | O | 2 | 2 | 6 |
| t | O | $\boxed{2}$ | $-2$ | O | $-2$ | $-1$ | O | 1 | 2 | ⟨1⟩ | O |
| P | O | O | 2 | $-M$ | $\boxed{1-3M}$ | $1+M/2$ | O | $3M/2$ | $1-M$ | | $5-3M$ |
| $x_1$ | 1 | O | 1 | O | $-1$ | O | O | 1 | 3 | NEG | 5 |
| s | O | O | O | 1 | $\boxed{3}$ | $\frac{1}{2}$ | 1 | $-\frac{1}{2}$ | 1 | ⟨1⟩ | 6 |
| $x_2$ | O | 1 | $-1$ | O | $-1$ | $-\frac{1}{2}$ | O | $\frac{1}{2}$ | 1 | NEG | O |
| P | O | O | 2 | $\boxed{-1/3}$ | O | $5/6+M$ | $-1/3+M$ | $1/6+M$ | $2/3$ | | $3+3M$ |
| $x_1$ | 1 | O | 1 | $1/3$ | O | $1/6$ | $1/3$ | $5/6$ | $9/3$ | 10 | 7 |
| $x_5$ | O | O | O | $\boxed{1/3}$ | 1 | $1/6$ | $1/3$ | $-1/6$ | $1/3$ | ⟨1⟩ | 2 |
| $x_2$ | O | 1 | $-1$ | $1/3$ | O | $-1/3$ | $1/3$ | $1/3$ | $4/3$ | 4 | 2 |
| P | O | O | 2 | O | 1 | $1+M$ | M | M | 1 | | $5+3M$ |
| $x_1$ | 1 | O | 1 | O | $-1$ | O | O | 1 | 3 | | 5 |
| $x_4$ | O | O | O | 1 | 3 | 2 | 1 | $-2$ | 1 | | 6 |
| $x_2$ | O | 1 | $-1$ | O | $-1$ | $-\frac{1}{2}$ | O | $\frac{1}{2}$ | 1 | | O |

This last tableau gives us a solution $x_1 = 3$, $x_2 = 1$, $x_3 = 0$, $x_4 = 1$, $x_5 = 0$ corresponding to $P = 1$.

This method of priming a problem is sometimes more lengthy than other ad hoc methods, but it has the benefit that it can be applied quite generally.

4.3(a)  This problem is in standard form with respect to the slacks r, s, t; the working is shown in Table S10.

## TABLE S10

| | x | y | z | r | s | t | | Check | | |
|---|---|---|---|---|---|---|---|---|---|---|
| P | (-1) | -1 | -1 | 0 | 0 | 0 | 0 | | -3 | choice of pivot column |
| r | [2] | -3 | 2 | 1 | 0 | 0 | 2 | (1) | 4 | |
| s | -3 | 2 | 2 | 0 | 1 | 0 | 2 | NEG | 4 | choice of pivot row. |
| t | 2 | 2 | -3 | 0 | 0 | 1 | 2 | 1 | 4 | |
| P | 0 | (-5/2) | 0 | ½ | 0 | 0 | 1 | | -1 | zero shows that to bring z into basis would not change profit |
| x | 1 | -3/2 | 1 | ½ | 0 | 0 | 1 | NEG | 2 | |
| s | 0 | -5/2 | 5 | 3/2 | 1 | 0 | 5 | NEG | 10 | |
| t | 0 | [5] | -5 | -1 | 0 | 1 | 0 | (0) | 0 | degeneracy, since t = 0 as well as y, z and r. |
| P | 0 | 0 | (-5/2) | 0 | 0 | ½ | 1 | | -1 | introduction of r does not change profit |
| x | 1 | 0 | -½ | 1/5 | 0 | 3/10 | 1 | NEG | 2 | |
| s | 0 | 0 | [5/2] | 1 | 1 | ½ | 5 | (2) | 10 | degeneracy, since y = z = r = t = 0 |
| y | 0 | 1 | -1 | -1/5 | 0 | 1/5 | 0 | NEG | 0 | 0/-1 = -0 |
| P | 0 | 0 | 0 | 1 | 1 | 1 | 6 | | 9 | |
| x | 1 | 0 | 0 | 2/5 | 1/5 | 2/5 | 2 | | 4 | |
| z | 0 | 0 | 1 | 2/5 | 2/5 | 1/5 | 2 | | 4 | solution not degenerate |
| y | 0 | 1 | 0 | 1/5 | 2/5 | 2/5 | 2 | | 4 | |

(b)  The problem is in standard form with respect to the slacks r, s, t and u, and the complete working is shown in Table S11.

## TABLE S11

| | x | y | r | s | t | u | | Check | | |
|---|---|---|---|---|---|---|---|---|---|---|
| P | -10 | (-11) | 0 | 0 | 0 | 0 | 0 | | -21 | |
| r | 1 | [10] | 1 | 0 | 0 | 0 | 100 | (10) | 112 | |
| s | 1 | 5 | 0 | 1 | 0 | 0 | 50 | 10 | 57 | choice of pivot row |
| t | 1 | 2 | 0 | 0 | 1 | 0 | 20 | 10 | 24 | |
| u | 1 | 0 | 0 | 0 | 0 | 1 | 10 | ∞ | 12 | no information |
| P | (-89/10) | 0 | 11/10 | 0 | 0 | 0 | 110 | | 102 1/5 | |
| y | 1/10 | 1 | 1/10 | 0 | 0 | 0 | 10 | 100 | 11 1/5 | |
| s | [½] | 0 | -½ | 1 | 0 | 0 | 0 | (0) | 1 | choice of pivot rows |
| t | 4/5 | 0 | -1/5 | 0 | 1 | 0 | 0 | 0 | 1 3/5 | doubly degenerate solution |
| u | 1 | 0 | 0 | 0 | 0 | 1 | 10 | 10 | 12 | |
| P | 0 | 0 | (-78/10) | 89/5 | 0 | 0 | 110 | | 120 | profit not increased |
| y | 0 | 1 | 1/5 | -1/5 | 0 | 0 | 10 | 50 | 11 | |
| x | 1 | 0 | -1 | 2 | 0 | 0 | 0 | NEG | 2 | -0 doubly degenerate solution |
| t | 0 | 0 | [3/5] | -8/5 | 1 | 0 | 0 | (0) | 0 | +0 |
| u | 0 | 0 | 1 | -2 | 0 | 1 | 10 | 10 | 10 | |
| P | 0 | 0 | 0 | (-3) | 13 | 0 | 110 | | 120 | profit not increased |
| y | 0 | 1 | 0 | 1/3 | -1/3 | 0 | 10 | 30 | 11 | |
| x | 1 | 0 | 0 | -2/3 | 5/3 | 0 | 0 | NEG | 2 | doubly degenerate solution |
| r | 0 | 0 | 1 | -8/3 | 5/3 | 0 | 0 | NEG | 0 | |
| u | 0 | 0 | 0 | [2/3] | -5/3 | 1 | 10 | (15) | 10 | |
| P | 0 | 0 | 0 | 0 | 11/2 | 9/2 | 155 | | 165 | |
| y | 0 | 1 | 0 | 0 | ½ | -½ | 5 | | 6 | |
| x | 1 | 0 | 0 | 0 | 0 | 1 | 10 | | 12 | non-degenerate optimal solution |
| r | 0 | 0 | 1 | 0 | -5 | 4 | 40 | | 40 | |
| s | 0 | 0 | 0 | 1 | -5/2 | 3/2 | 15 | | 15 | |

4.4(a) The problem is not in standard form with respect to any basis, although it is nearly so: x = 10, s = 10, t = 8, y = 0, z = 0, r = 0 satisfies the constraints but x is included in the cost function. Hence if we substitute for x from the first constraint the problem will be in standard form with respect to this solution. The cost function is then

$$10 - y + 2z - 2r$$

Alternatively we could find an initial solution by means of dummy variables, but this requires extra working. The calculation is as shown in Table S12. The two optimal solutions are therefore seen to be

$$x = 2, \quad y = 0, \quad z = 0, \quad r = 4, \quad s = 1, \quad t = 0,$$

and $x = 4, \quad y = 1, \quad z = 0, \quad r = 3\frac{1}{2}, \quad s = 0, \quad t = 0,$

both having a cost of 2 units.

### TABLE S12

| | x | y | z | r | s | t | | chk |
|---|---|---|---|---|---|---|---|---|
| P | O | -1 | 2 | (-2) | O | O | -10 | -11 |
| x | 1 | -1 | 1 | 2 | O | O | 10 | 5 | 13 |
| s | O | 1 | -1 | O | 1 | O | 1 | ∞ | 2 |
| t | O | 1 | O | [2] | O | 1 | 8 | (4) | 12 |
| P | O | (O) | 2 | O | O | 1 | -2 | 1 |
| x | 1 | -2 | 1 | O | O | -1 | 2 | NEG | 1 |
| s | O | [1] | -1 | O | 1 | O | 1 | (1) | 2 |
| r | O | ½ | O | 1 | O | ½ | 4 | 8 | 6 |
| P | O | O | 2 | O | (O) | 1 | -2 | 1 |
| x | 1 | O | -1 | O | 2 | -1 | 4 | 5 |
| y | O | 1 | -1 | O | 1 | O | 1 | 2 |
| r | O | O | ½ | 1 | -½ | ½ | 3½ | 5 |

Notes

A solution, but this zero, not corresponding to a variable in the basis, shows that y may be brought in and the profit remain unaltered.

Another solution with the same profit. The existence of the other solution is indicated by the ringed zero.

(b) Inserting slacks, the problem becomes:

Maximise     $-3x - 2y - z$

subject to     $x + y + 2z + r = 6$

$x + 4y - s = 12$

$2x - y + 3z - t = 7$

$x - z - u = 2$

$y + 4z - v = 20$

$x,y,z,r,s,t,u,v \geq 0$

Subtracting each of the bracketed equations from the last (the one with the largest constant term) and adding a dummy variables gives:

Maximise $\quad$ -3x - 2y - z $\qquad\qquad$ -Mw

subject to $\quad$ x + y + 2z + r $\qquad\qquad\qquad$ = 6

$\qquad\qquad$ - x - 3y + 4z $\quad$ + s $\qquad$ - v $\quad$ = 8

$\qquad\qquad$ -2x + 2y + z $\qquad\quad$ + t $\quad$ - v $\quad$ = 13

$\qquad\qquad$ - x + y + 5z $\qquad\qquad$ + u - v $\quad$ = 18

$\qquad\qquad\qquad$ y + 4z $\qquad\qquad\qquad$ - v + w = 20

$\qquad\qquad$ x,y,r,s,t,u,v,w $\geqslant$ 0

To set the problem into standard form with respect to r=s=t=u=w=0 we must eliminate the w term in the profit function using the last constraint. It now becomes:

Maximise $\quad$ -3x + (M-2)y + (4M-1)z $\quad$ - Mv $\quad$ - 20M

The complete working is set out in Table S13.

### TABLE S13

| | x | y | z | r | s | t | u | v | w | | | chk |
|---|---|---|---|---|---|---|---|---|---|---|---|---|
| P | 3 | 2-M | (1-4M) | 0 | 0 | 0 | 0 | M | 0 | -20M | | 6-24M |
| r | 1 | 1 | 2 | 1 | 0 | 0 | 0 | 0 | 0 | 6 | 3 | 11 |
| s | -1 | -3 | [4] | 0 | 1 | 0 | 0 | -1 | 0 | 8 | (2) | 8 |
| t | -2 | 2 | 1 | 0 | 0 | 1 | 0 | -1 | 0 | 13 | 13 | 14 |
| u | -1 | 1 | 5 | 0 | 0 | 0 | 1 | -1 | 0 | 18 | 18/5 | 23 |
| w | 0 | 1 | 4 | 0 | 0 | 0 | 0 | -1 | 1 | 20 | 5 | 25 |
| P | 13/4-M | (11/4-4M) | 0 | 0 | M-1/4 | 0 | 0 | 1/4 | 0 | -12M-2 | | 4-16M |
| r | 3/2 | [5/2] | 0 | 1 | -1/2 | 0 | 0 | 1/2 | 0 | 2 | (4/5) | 7 |
| z | -1/4 | -3/4 | 1 | 0 | 1/4 | 0 | 0 | -1/4 | 0 | 2 | NEG | 2 |
| t | -7/4 | 11/4 | 0 | 0 | -1/4 | 1 | 0 | -3/4 | 0 | 11 | 4 | 12 |
| u | 1/4 | 19/4 | 0 | 0 | -5/4 | 0 | 1 | 1/4 | 0 | 8 | 32/19 | 13 |
| w | 1 | 4 | 0 | 0 | -1 | 0 | 0 | 0 | 1 | 12 | 3 | 17 |
| P | 8/5+7M | 0 | 0 | 8M/5-11/10 | M/5+3/100 | 0 | 0 | 4M/5-3/10 | 0 | -21/5-44M/5 | | -37/10-24M/5 |
| y | 3/5 | 1 | 0 | 2/5 | -1/5 | 0 | 0 | 1/5 | 0 | 4/5 | | 14/5 |
| z | 1/5 | 0 | 1 | 3/10 | 1/10 | 0 | 0 | -1/10 | 0 | 13/5 | | 41/10 |
| t | -17/5 | 0 | 0 | -11/10 | 3/10 | 1 | 0 | -13/10 | 0 | 44/5 | | 43/10 |
| u | -13/5 | 0 | 0 | -19/10 | -3/10 | 0 | 1 | -7/10 | 0 | 21/5 | | -3/10 |
| w | -7/5 | 0 | 0 | -12/5 | -1/5 | 0 | 0 | -4/5 | 1 | 48/5 | | 29/5 |

All elements are non-negative, but the dummy variable w is still part of the solution. Hence no feasible solution to original problem exists.

(c) This problem is in standard form with respect to the obvious initial solution using the slacks r, s, t (Table S14).

The division column shows that we can increase y without limit, hence giving an infinite solution to the problem. (An inspection of the original equations shows this to be true.)

TABLE S14

| | x | y | z | r | z | t | | | Chk |
|---|---|---|---|---|---|---|---|---|---|
| P | (-2) | -1 | 1 | 0 | 0 | 0 | 0 | | -2 |
| r | 0 | -1 | 1 | 1 | 0 | 0 | 4 | ∞ | 5 |
| s | [3] | 0 | 4 | 0 | 1 | 0 | 6 | (2) | 14 |
| t | -1 | -2 | 3 | 0 | 0 | 1 | 8 | Neg | 9 |
| P | 0 | (-1) | $^{11}/_3$ | 0 | $^2/_3$ | 0 | 4 | | $^{22}/_3$ |
| r | 0 | -1 | 1 | 1 | 0 | 0 | 4 | Neg | 5 |
| x | 1 | 0 | $^4/_3$ | 0 | $^1/_3$ | 0 | 2 | ∞ | $^{14}/_3$ |
| t | 0 | -2 | $^{13}/_3$ | 0 | $^1/_3$ | 1 | 10 | Neg | $^{41}/_3$ |

4.5(a)  The solution is given in Table S15.

TABLE S15

Knife settings

| Width | 1 | 2 | 3 | 4 | 5 |
|---|---|---|---|---|---|
| 1.20 | 1 | 1 | 0 | 0 | 0 |
| 0.80 | 1 | 0 | 2 | 1 | 0 |
| 0.45 | 0 | 2 | 1 | 3 | 5 |
| Loss | .30 | .20 | .25 | .15 | .05 |

(b)  Minimise  $0.3x_1 + 0.2x_2 + 0.25x_3 + 0.15x_4 + 0.05x_5$

subject to

$$x_1 + x_2 \geqslant 30$$
$$x_1 + 2x_3 + x_4 \geqslant 40$$
$$2x_2 + x_3 + 3x_4 + 5x_5 \geqslant 50$$
$$x_1 \ldots \ldots x_5 \geqslant 0$$

The addition of dummy variables allows us to find an initial solution.  This leads straightforwardly to

$x_1 = 26$, $x_2 = 4$, $x_3 = 0$, $x_4 = 14$, $x_5 = 0$

It is very convenient that this is an integer solution: usually this would not be the case and so approximations would have to be made.

5.1

TABLE S16

| From Depots: | | $D_1$ | $D_2$ | $D_3$ | Orders |
|---|---|---|---|---|---|
| To Retailers: | $R_1$ | 17 | 16 | 14 | 52 |
| | $R_3$ | 15 | 11 | 14 | 35 |
| | $R_4$ | 12 | 11 | 10 | 41 |
| Stocks | | 43 | 55 | 30 | 128 |

The unit costs are as shown in Table S16.

Using the notation of the text, the problem can be written as follows:

Minimise $\quad C = 17x_1 + 16y_1 + 14z_1 + 15x_3 + 11y_3 + 14z_3 + 12x_4 + 11y_4 + 10z_4$

subject to

$$x_1 + x_3 + x_4 \qquad\qquad\qquad = 43$$
$$y_1 + y_2 + y_4 \qquad\qquad = 55$$
$$z_1 + z_3 + z_4 = 30$$
$$x_1 \qquad + y_1 \qquad + z_1 \qquad = 52$$
$$x_3 \qquad + y_3 \qquad + z_3 \qquad = 35$$
$$x_4 \qquad + y_4 \qquad + z_4 = 41$$

$$x_1, x_3, x_4, y_1, y_3, y_4, z_1, z_3, z_4, \geqslant 0$$

Only 5 of these constraints are independent, so that for a feasible solution we must have just 5 non-zero variables. Such a solution can be obtained by using the cheapest routes first as in Table S17.

*TABLE* S17

| Variables: | x | y | z | Totals |
|---|---|---|---|---|
| Subscripts: 1 | 43 <br> 17 | 9 <br> 16 | 0 <br> 14 | 52 |
| 3 | 0 <br> 15 | 35 <br> 11 | 0 <br> 14 | 35 |
| 4 | 0 <br> 12 | 11 <br> 11 | 30 <br> 10 | 41 |
| Totals | 43 | 55 | 30 | 128 |

The corresponding cost is 1681 units.

5.2      Increase $x_4$ by $\theta$. In order to keep the column 1 total at 43 (Table S18) we must subtract $\theta$ from $x_1$. We must then add $\theta$ to $y_1$ (not to $z_1$, otherwise we should also be bringing $z_1$ into the basis) and finally subtract $\theta$ from $y_4$. The marginal totals are all unchanged. The positivity constraints give $43 - \theta \geqslant 0$, $9 + \theta \geqslant 0$, $\theta \geqslant 0$, $11 - \theta \geqslant 0$. This shows that the maximum allowable value for $\theta$ is 11, corresponding to $y_4$ leaving the basis. The new basic feasible solution is as shown in Table S19 (where the unit costs have been omitted). The corresponding total cost is unchanged.

1

TABLE S18

| | x | y | z | Totals |
|---|---|---|---|---|
| 1 | 43-θ __17__ | 9+θ __16__ | O __14__ | 52 |
| 3 | O __15__ | 35 __11__ | O __14__ | 35 |
| 4 | θ __12__ | 11-θ __11__ | 30 __10__ | 4i |
| Totals | 43 | 55 | 3O | 128 |

TABLE S19

| | x | y | z | |
|---|---|---|---|---|
| 1 | 32 | 20 | O | 52 |
| 3 | O | 35 | O | 35 |
| 4 | 11 | O | 30 | 41 |
| | 43 | 55 | 30 | 128 |

5.3

$x_3$:                                    $x_4$:

Coefficients:   $15 - 17 + 16 - 11 = 3$  :  $12 - 17 + 16 - 11 = 0$

$z_1$:                                    $z_2$:

Coefficients:   $14 - 10 + 11 - 16 = -1$ :  $14 - 11 + 11 - 10 = 4$

Fig. S.21

Consider Fig.S21.   (The direction of the arrows is unimportant.)
The introduction of $x_4$ would have no effect on the cost, since its
coefficient is zero;  however $z_1$ could be introduced to advantage,
since its coefficient is -1.

5.4   As in Table S2O, add θ to $z_1$ and $y_4$, subtract θ from $y_1$ and $z_4$,
following the circuit of Solution 5.3.   The non-negativity

constraints give $\theta \geqslant 0$, $30 - \theta \geqslant 0$, $11 + \theta \geqslant 0$, $9 - \theta \geqslant 0$, showing that we cannot increase $\theta$ by more than 9 (Table S21).

*TABLE* S20

|   | x | y | z |   |
|---|---|---|---|---|
| 1 | 43 <br> 17 | 9−θ <br> 16 | 0+θ <br> 14 | 52 |
| 3 | 0 <br> 15 | 35 <br> 11 | 0 <br> 14 | 35 |
| 4 | 0 <br> 12 | 11+θ <br> 11 | 30−θ <br> 10 | 41 |
|   | 43 | 55 | 30 | 128 |

*TABLE* S21

|   | x | y | z |   |
|---|---|---|---|---|
| 1 | 43 <br> 17 | 0 <br> 16 | 9 <br> 14 | 52 |
| 3 | 0 <br> 15 | 35 <br> 11 | 0 <br> 14 | 35 |
| 4 | 0 <br> 12 | 20 <br> 11 | 21 <br> 10 | 41 |
|   | 43 | 55 | 30 | 128 |

We must now calculate the coefficients of the cost function by following suitable closed circuits and using the unit costs:

$x_3$ : $15 - 17 + 14 - 10 + 11 - 11 = 2$

$x_4$ : $12 - 17 + 14 - 10$ $= -1$ : introduction of $x_4$ will reduce costs

$y_1$ : $16 - 14 + 10 - 11$ $= 1$

$z_3$ : $14 - 11 + 11 - 10$ $= 4$

*TABLE* S22

|   | x | y | z |   |
|---|---|---|---|---|
| 1 | 43−θ <br> 17 | 0 <br> 16 | 9+θ <br> 14 | 52 |
| 3 | 0 <br> 15 | 35 <br> 11 | 0 <br> 14 | 35 |
| 4 | θ <br> 12 | 20 <br> 11 | 21−θ <br> 10 | 41 |
|   | 43 | 55 | 30 | 128 |

TABLE S23

| | x | y | z | |
|---|---|---|---|---|
| 1 | 22 <sub>17</sub> | O <sub>16</sub> | 30 <sub>14</sub> | 52 |
| 3 | O <sub>15</sub> | 35 <sub>11</sub> | O <sub>14</sub> | 35 |
| 4 | 21 <sub>12</sub> | 20 <sub>11</sub> | O <sub>10</sub> | 41 |
| | 43 | 55 | 30 | 128 |

From Table S22 we see that $\theta$ = 21 is the maximum value for $x_4$. This in turn leads to Table S23.   The cost coefficients must now be calculated:

$x_3 = 15 - 11 + 11 - 12$      $= 3$

$y_1 = 16 - 11 + 12 - 17$      $= 0$ : introduction of $y_1$ will not

$z_3 = 14 - 14 + 17 - 12 + 11 - 11 = 5$      change the cost

$z_4 = 10 - 14 + 17 - 12$      $= 1$

Since none of these coefficients is negative,  an optimal solution has been obtained, the cost being 1651 units.

For an alternative optimal solution introduce $y_1$ (Table S24):  with $\theta$ = 20 we obtain Table S25.

TABLE S24

| | x | y | z | |
|---|---|---|---|---|
| 1 | $22-\theta$ <sub>17</sub> | $\theta$ <sub>16</sub> | 30 <sub>14</sub> | 52 |
| 3 | O <sub>15</sub> | 35 <sub>11</sub> | O <sub>14</sub> | 35 |
| 4 | $21+\theta$ <sub>12</sub> | $20+\theta$ <sub>11</sub> | O <sub>10</sub> | 41 |
| | 43 | 55 | 30 | 128 |

### TABLE S25

|   | x |   | y |   | z |   |   |
|---|---|---|---|---|---|---|---|
| 1 | 2 | 17 | 20 | 16 | 30 | 14 | 52 |
| 3 | 0 | 15 | 35 | 11 | 0 | 14 | 35 |
| 4 | 41 | 12 | 0 | 11 | 0 | 10 | 41 |
|   | 43 |   | 55 |   | 30 |   | 128 |

The cost coefficients are:

$x_3 = 15 - 17 + 16 - 11 = 3$

$y_4 = 11 - 12 + 17 - 16 = 0$ : introduction of $y_4$ returns us to the previous solution

$z_3 = 14 - 14 + 16 - 11 = 5$

$z_4 = 10 - 12 + 17 - 14 = 1$

This is an alternative solution, again with a cost of 1651 units.

### TABLE S26

5.5

| | Shadow Costs | x 17 | y 16 | z 15 | |
|---|---|---|---|---|---|
| 1 | 0 | 43 (17) | 9−θ (16) | 0+θ (14) ⊖(-1) | 52 |
| 3 | −5 | 3   0 (15) | 35 (11) | 4   0 (14) | 35 |
| 4 | −5 | 0   0 (12) | 11+θ (11) | 30−θ (10) | 41 |
| | | 43 | 55 | 30 | 128 |

The initial basic feasible solution of Solution 5.1 is displayed in Table S26 together with shadow costs obtained from the routes used, having arbitrarily chosen that for retailer 1 to be zero. The costs are next calculated for the unused routes. For example, $x_3$ has the value $15 - (17 - 5) = 3$. Introducing $z_1$ into the basis as before (Solution 5.4), we find that no more than 9 may be sent by this route. The next tableau is as shown in Table S27. Again the value of zero for the shadow cost of retailer 1 is assumed and the others are then uniquely defined. The cost coefficients corresponding to the variables set to zero are then calculated and from these it can be seen that $x_4$ should be brought into the basis — with a value of 21 (see Table S28).

## TABLE S27

| | Shadow Costs | x 17 | y 15 | z 14 | |
|---|---|---|---|---|---|
| 1 | 0 | ¹ 43−θ ⟨17⟩ | 0 ⟨16⟩ | 9+θ ⟨14⟩ | 52 |
| 3 | −4 | ² 0 ⟨15⟩ | 35 ⟨11⟩ | ⁴ 0 ⟨14⟩ | 35 |
| 4 | −4 | ⟨-1⟩ 0+θ ⟨12⟩ | 20 ⟨11⟩ | 21−θ ⟨10⟩ | 41 |
| | | 43 | 55 | 30 | 128 |

## TABLE S28

| | Shadow Costs | x 17 | y 16 | z 14 | |
|---|---|---|---|---|---|
| 1 | 0 | 22−θ ⟨17⟩ | ⟨0⟩ 0+θ ⟨16⟩ | 30 ⟨14⟩ | 52 |
| 3 | −5 | ³ 0 ⟨15⟩ | 35 ⟨11⟩ | ⁵ 0 ⟨14⟩ | 35 |
| 4 | −5 | 21+θ ⟨12⟩ | 20−θ ⟨11⟩ | ¹ 0 ⟨10⟩ | 41 |
| | | 43 | 55 | 30 | 128 |

A solution has been obtained.   The zero indicates another solution given by θ = 20.

5.6  As there are only 4 non-zero routes the problem is degenerate (Table S29).

## TABLE S29

| | x | y | z | |
|---|---|---|---|---|
| 1 | 43 ⟨17⟩ | 0 ⟨16⟩ | 0 ⟨14⟩ | 43 |
| 3 | 0 ⟨15⟩ | 35 ⟨11⟩ | 0 ⟨14⟩ | 35 |
| 4 | 0 ⟨12⟩ | 11 ⟨11⟩ | 30 ⟨10⟩ | 41 |
| | 43 | 46 | 30 | 119 |

*TABLE* S30

|  | x | y | z |  |
|---|---|---|---|---|
| 1 | 43 <br> 17 | O <br> 16 | O <br> 14 | 43 |
| 3 | ε <br> 15 | 35−ε <br> 11 | O <br> 14 | 35 |
| 4 | O <br> 12 | 11+2ε <br> 11 | 30+ε <br> 10 | 41+3ε |
|  | 43+ε | 46+ε | 30+ε | 119+3ε |

The insertion of ε's and the consequent basic feasible solution are shown in Table S30.  We can now add shadow costs and proceed as usual (Table S31) with $\theta = \varepsilon$.

We see that the solution would still be degenerate if it were not for ε, hence it cannot yet be discarded.  The next step in the solution is shown in Table S32 with $\theta = 30 + \varepsilon$.

*TABLE* S31

| | Shadow Costs | x <br> 17 | y <br> 13 | z <br> 12 | |
|---|---|---|---|---|---|
| 1 | O | 43 <br> 17 | O <br> 16 | O <br> 14 | 43 |
| 3 | −2 | ε−θ <br> 15 | 35−ε+θ <br> 11 | O <br> 14 | 35 |
| 4 | −2 | O+θ <br> 12 | 11+2ε−θ <br> 11 | 30+ε <br> 10 | 41+3ε |
| | | 43+ε | 46+ε | 30+ε | 119+3ε |

The ε's can now be dropped since the solution is non-degenerate. The optimal solution has been obtained, as can be seen from Table S33.  An alternative can be found by bringing $y_1$ into the basis with a value of 11(+ε).

## TABLE S32

| | Shadow Costs | x — 17 | y — 16 | z — 15 | |
|---|---|---|---|---|---|
| 1 | O | $43-\theta$ (0) · 17 | O · 16 | $0+\theta$ (-1) · 14 | 43 |
| 3 | -5 | O (3) · 15 | 35 (4) · 11 | O · 14 | 35 |
| 4 | -5 | $\varepsilon+\theta$ · 12 | $11+\varepsilon$ · 11 | $30+\varepsilon-\theta$ · 10 | $41+3\varepsilon$ |
| | | $43+\varepsilon$ | $46+\varepsilon$ | $30+\varepsilon$ | $119+3\varepsilon$ |

## TABLE S33

| | Shadow Costs | x — 17 | y — 16 | z — 14 | |
|---|---|---|---|---|---|
| 1 | O | $13-\varepsilon-\theta$ (0) · 17 | $0+\theta$ · 16 | $30+\varepsilon$ · 14 | 43 |
| 3 | -5 | O (3) · 15 | 35 · 11 | O (5) · 14 | 35 |
| 4 | -5 | $30+2\varepsilon+\theta$ · 12 | $11+\varepsilon-\theta$ · 11 | O (1) · 10 | $41+3\varepsilon$ |
| | | $43+\varepsilon$ | $46+\varepsilon$ | $30+\varepsilon$ | $119+3\varepsilon$ |

5.7(a)  The problem is out of balance, in that the orders total 128 but there are 135 available.  We must therefore add a slack retailer showing zero unit transport costs (Table S34).

Using the least cost routes to find an initial solution, we see that it is basic and non-degenerate (6 non-zero routes).  The working is shown in Tables S35 to S37.

### TABLE S34 : $\theta = 7$

| Shadow Costs | 17 | 16 | 15 | Orders |
|---|---|---|---|---|
| O | 50-θ   [17] | 2+θ   [16] | ⁻¹ O   [14] | 52 |
| -5 | ³ O   [15] | 35   [11] | ⁴ O   [14] | 35 |
| -5 | ⁰ O   [12] | 18-θ   [11] | 23+θ   [10] | 41 |
| Slack  -15 | ⁻² O+θ   [0] | ⁻¹ O   [0] | 7-θ   [0] | 7 |
| Stocks | 50 | 55 | 30 | 135 |

### TABLE S35 : $\theta = 9$

| Shadow Costs | 17 | 16 | 15 | Orders |
|---|---|---|---|---|
| O | 43   [17] | 9-θ   [16] | ⊖ O+θ   [14] | 52 |
| -5 | ³ O   [15] | 35   [11] | ⁴ O   [14] | 35 |
| -5 | ⁰ O   [12] | 11+θ   [11] | 30-θ   [10] | 41 |
| -17 | 7   [0] | ¹ O   [0] | ² O   [0] | 7 |
| Stocks | 50 | 55 | 30 | 135 |

## TABLE S36 : θ = 21

| Shadow Costs | 17 | 15 | 14 | Orders |
|---|---|---|---|---|
| O | [1] 43−θ (17) | O (16) | 9+θ (14) | 52 |
| −4 | [2] O (15) | [4] 35 (11) | O (14) | 35 |
| −4 | (−1) O+θ (12) | 20 (11) | 21−θ (10) | 41 |
| −17 | [2] 7 (0) | [3] O (0) | O (0) | 7 |
| Stocks | 50 | 55 | 30 | 135 |

## TABLE S37 : θ = 20

| Shadow Costs | 17 | 16 | 14 | Orders |
|---|---|---|---|---|
| O | 22−θ (17) | ⓪ O+θ (16) | 30 (14) | 52 |
| −5 | [3] O (15) | 35 (11) | [5] O (14) | 35 |
| −5 | 21+θ (12) | 20−θ (11) | [1] O (10) | 41 |
| −17 | 7 (0) | [1] O (0) | [3] O (0) | 7 |
| Stocks | 50 | 55 | 30 | 135 |

The last tableau gives an optimal solution: an alternative can be obtained by taking θ = 20.

(b) This problem is also out of balance, requiring a slack distributor with zero associated unit costs (Table S38).

### TABLE S38 : θ = 3

| Shadow Costs | 17 | 16 | 15 | 5 | Orders |
|---|---|---|---|---|---|
| 0 | 40 _17_ | 12−θ _16_ | −1 · 0 _14_ | ⊝5 · 0+θ _0_ | 52 |
| −5 | 3 · 0 _15_ | 35 _11_ | 4 · 0 _14_ | 0 · 0 _0_ | 35 |
| −5 | 0 · 0 _12_ | 8+θ _11_ | 30 _10_ | 3−θ _0_ | 41 |
| Stocks | 40 | 55 | 30 | 3 | 128 |

### TABLE S39 : θ = 9

| Shadow Costs | 17 | 16 | 15 | 0 | Orders |
|---|---|---|---|---|---|
| 0 | 40 _17_ | 9−θ _16_ | ⊝1 · 0+θ _14_ | 3 _0_ | 52 |
| −5 | 3 · 0 _15_ | 35 _11_ | 4 · 0 _14_ | 5 · 0 _0_ | 35 |
| −5 | 0 · 0 _12_ | 11+θ _11_ | 30−θ _10_ | 5 · 0 _0_ | 41 |
| Stocks | 40 | 55 | 30 | 3 | 128 |

### TABLE S40 : θ = 21

| Shadow Costs | 17 | 15 | 14 | 0 | Orders |
|---|---|---|---|---|---|
| 0 | 40−θ _17_ | 1 · 0 _16_ | 9+θ _14_ | 3 _0_ | 52 |
| −4 | 2 · 0 _15_ | 35 _11_ | 4 · 0 _14_ | 4 · 0 _0_ | 35 |
| −4 | ⊝1 · 0+θ _12_ | 20 _11_ | 21−θ _10_ | 4 · 0 _0_ | 41 |
| Stocks | 40 | 55 | 30 | 3 | 128 |

## TABLE S41

| Shadow Costs | 17 | 16 | 14 | 0 | Orders |
|---|---|---|---|---|---|
| 0 | 19-θ <br> 17 | 0 <br> 0+θ <br> 16 | 30 <br> 14 | 3 <br> 0 | 52 |
| -5 | 3 <br> 0 <br> 15 | 35 <br> 11 | 5 <br> 0 <br> 14 | 5 <br> 0 <br> 0 | 35 |
| -5 | 21+θ <br> 12 | 20-θ <br> 11 | 1 <br> 0 <br> 10 | 5 <br> 0 <br> 0 | 41 |
| Stocks | 40 | 55 | 30 | 3 | 128 |

The working is shown in Tables S38 to S41. The last tableau gives an optimal solution: the alternative is found by taking $\theta = 19$.

5.8(a)

## TABLE S42

| From | Shadow Costs | To 1 <br> 4.0 | 2 <br> 3.3 | 4 <br> -0.7 | 7 <br> -3.8 | 8 <br> -5.0 | 9 <br> -1.8 | 10 <br> -2.1 | 11 <br> -3.3 | 12 <br> -1.1 | Totals |
|---|---|---|---|---|---|---|---|---|---|---|---|---|
| 3 | 0 | 5 <br> 4.0 | 1.8 <br> 0 <br> 5.1 | 5.8 <br> 0 <br> 5.1 | 13.2 <br> 0 <br> 9.4 | 13.4 <br> 0 <br> 8.4 | 13.9 <br> 0 <br> 12.1 | 13.8 <br> 0 <br> 11.7 | 14.9 <br> 0 <br> 11.6 | 14.6 <br> 0 <br> 13.5 | 5 |
| 5 | 6.7 | 15 <br> 10.7 | 70 <br> 10.0 | 18 <br> 6.0 | 2.1 <br> 0 <br> 5.0 | ⓪ <br> 0+θ <br> 1.7 | 0.1 <br> 0 <br> 5.0 | 0.7 <br> 0 <br> 5.3 | 1.0 <br> 0 <br> 4.4 | 213-θ <br> 5.6 | 316 |
| 6 | 6.2 | 1.4 <br> 0 <br> 11.6 | 0.6 <br> 0 <br> 10.1 | 0.9 <br> 0 <br> 6.4 | 31 <br> 2.4 | 52-θ <br> 1.2 | 15 <br> 4.4 | 161 <br> 4.1 | 152 <br> 2.9 | 191+θ <br> 5.1 | 602 |
| Totals | | 20 | 70 | 18 | 31 | 52 | 15 | 161 | 152 | 404 | 923 |

An initial basic feasible solution (11 non-zero entries) is shown in Table S42, using the 'least cost' method. It is seen to be optimal although there exists another optimal solution given by $\theta = 52$.

(b) (i) If this result is not intuitively obvious, trial and error should convince you.

(ii) The problem can now be written as follows:

$$\text{Minimise} \quad 15x_{11} + 14x_{12} + 14x_{13} + 17x_{14} + 16x_{21} + 17x_{22} + \ldots\ldots\ldots\ldots + 11x_{43} + 10x_{44}$$

subject to

$$
\begin{aligned}
x_{11} + x_{12} + x_{13} + x_{14} &= 1 \\
x_{21} + x_{22} + x_{23} + x_{24} &= 1 \\
x_{31} + x_{32} + x_{33} + x_{34} &= 1 \\
x_{41} + x_{42} + x_{43} + x_{44} &= 1 \\
x_{11} \qquad\qquad + x_{21} \qquad\qquad + x_{31} \qquad\qquad + x_{41} &= 1 \\
x_{12} \qquad\qquad + x_{22} \qquad\qquad + x_{32} \qquad\qquad + x_{42} &= 1 \\
x_{13} \qquad\qquad + x_{23} \qquad\qquad + x_{33} \qquad\qquad + x_{43} &= 1 \\
x_{14} \qquad\qquad + x_{24} \qquad\qquad + x_{34} \qquad\qquad + x_{44} &= 1 \\
x_{11} \ldots x_{44} &\geqslant 0
\end{aligned}
$$

The constraints will ensure the solution has the required properties. An initial 'minimum-cost' solution is shown in Table S43.

### TABLE S43

|  | $P_1$ | $P_2$ | $P_3$ | $P_4$ |  |
|---|---|---|---|---|---|
| $J_1$ | 0 _(15)_ | 0 _(14)_ | 1 _(14)_ | 0 _(17)_ | 1 |
| $J_2$ | 1 _(16)_ | 0 _(17)_ | 0 _(18)_ | 0 _(17)_ | 1 |
| $J_3$ | 0 _(12)_ | 0 _(11)_ | 0 _(10)_ | 1 _(9)_ | 1 |
| $J_4$ | 0 _(10)_ | 1 _(8)_ | 0 _(11)_ | 0 _(10)_ | 1 |
|  | 1 | 1 | 1 | 1 | 4 |

This involves only 4 non-zero 'routes', hence it is a degenerate solution and we must use the '$\varepsilon$' technique to find an initial solution. The following two tableaux (Tables S44, S45) show that the optimal solution is given by allocating $J_1$ to $P_3$, $J_2$ to $P_1$, $J_3$ to $P_4$ and $J_4$ to $P_2$, costing the company 47 units.

6.1(a)  (i) and (ii) are competitions in the usual sense of the word, each participant being a competitor.

(iii)  is a competition in a more specialized sense, the two 'sides' being the competitors (to a first approximation).

(iv)  A competition, but there seems to be only one competitor, unless we consider "nature" to be the oponent.

(v)  Not usually thought of as a competition (or else we might have even more divorces).

(vi)  A competition against nature?

(vii)  This course is designed not to be a competition. Even examinations are not competitive since everyone *can* pass and

## TABLE S44

| | Shadow Costs | P$_1$ 12 | P$_2$ 10 | P$_3$ 14 | P$_4$ 12 | |
|---|---|---|---|---|---|---|
| J$_1$ | 0 | [3] O — 15 | [4] O — 14 | 1 — 14 | [5] O — 17 | 1 |
| J$_2$ | 4 | 1−ε+θ — 16 | [3] O — 17 | ε−θ — 18 | [1] O — 17 | 1 |
| J$_3$ | −3 | [3] O — 12 | [4] O — 11 | [−1] O — 10 | 1 — 9 | 1 |
| J$_4$ | −2 | 2ε−θ — 10 | 1+ε — 8 | [⊖1] O+θ — 11 | ε — 10 | 1+4ε |
| | | 1+ε | 1+ε | 1+ε | 1+ε | 4+4ε |

## TABLE S45

| | Shadow Costs | P$_1$ 13 | P$_2$ 11 | P$_3$ 14 | P$_4$ 13 | |
|---|---|---|---|---|---|---|
| J$_1$ | 0 | [2] O — 15 | [3] O — 14 | 1 — 14 | [4] O — 17 | 1 |
| J$_2$ | 3 | 1 — 16 | [3] O — 17 | [1] O — 18 | [1] O — 17 | 1 |
| J$_3$ | −4 | [3] O — 12 | [4] O — 11 | [0] O — 10 | 1 — 9 | 1 |
| J$_4$ | −3 | ε — 10 | 1+ε — 8 | ε — 11 | ε — 10 | 1+4ε |
| | | 1+ε | 1+ε | 1+ε | 1+ε | 4+4ε |

should do so if they follow the course diligently. The *outcome* of the exams, however, will help in the undoubted competition for jobs – unless you want to 'drop-out'.

(N.B. The results (v) and (vii) are rather subjective and you may disagree with them completely.)

(b) (i) Two players (by our specialized definition).

(ii) The number of horses entered in the race.

(iii) Two players, the opposing sides. Although rivalry existed between the allies, they had a common goal.

(iv) If run on strictly party political lines, then probably only about half a dozen players.

6.2 (a)   (i)   Money, matchsticks.

(ii)   Power, influence, money.

(iii)   Satisfaction, freedom from boredom.

(iv)   Beer (if played in a pub).

(v)   The mind boggles!

(b)   They are all zero sum with the possible exception of (ii), party politics.   It is possible in a political situation that everybody 'wins', reaches their objectives (e.g. entry into the Common Market) or 'loses' (e.g. Second World War).

6.3   For A or B either   (a)   always write down 1,

or   (b)   always write down 2,

or   (c)   write 1, 2 alternately,

or   (d)   write 1, 2 at random,

or   (e)   write the same number as the opponent wrote last time,

etc.

6.4 (i)   See Table S46

TABLE S46

| A:<br>B | Shout 1 | Shout 2 |
|---|---|---|
| Shout 1 | 0 | 1 |
| Shout 2 | -1 | -3 |

(ii) $V(B) = 0$   (B shouts 1, A shouts 1).

(iii) $V(A) = -V(B)$ since A's gains are B's losses and vice versa.

6.5   (i)   If A plays $A_1$ he will win not less than 1.

If A plays $A_2$ he will win not less than -6.

So he will play $A_1$ and win at least 1.

(ii)   If B plays $B_1$ he will lose not more than 1.

If B plays $B_2$ he will lose not more than 4.

If B plays $B_3$ he will lose not more than 2.

So B will  play $B_1$ and lose not more than 1.

(iii)   Since these two values are equal, the value of the game is 1 to A.

(iv)   The game is not fair since $V(A) = 1$, not zero.

(v)   This game is in equilibrium since

(a) If B plays his optimum strategy $B_1$, A can only lose by moving from $A_1$ (0 instead of 1).

(b) If A plays $A_1$, but B does not play $B_1$, he will lose 2 or 3 instead of 1.

6.6    The game is as shown in Table S47

### TABLE S47

| B:<br>A | $B_1$ | $B_2$ | $B_3$ | Row Min | |
|---|---|---|---|---|---|
| $A_1$ | 2 | 4 | 6 | 2 | ← maximin |
| $A_2$ | -2 | -4 | -6 | -6 | |
| $A_3$ | 0 | -2 | -4 | -4 | |
| Col Max | 2 | 4 | 6 | | |

minimax

(i)    2, $(A_1)$

(ii)   2, $(B_1)$

(iii)  2, $(A_1, B_1)$

(iv)   2

(v)    $A_1$

(vi)   $B_1$

6.7    No, since the moves are made simultaneously.

6.8    (i)    See Table S48.

### TABLE S48

| B:<br>A | H | T | Row Min |
|---|---|---|---|
| H | -1 | 1 | -1 |
| T | 1 | -1 | -1 |
| Col Max | 1 | 1 | |

(ii)   Minimax value = 1

       Maximin value = -1

(iii)  No, since minimax ≠ maximin.

(iv)   No, since there is no saddle point.

6.9    (i)    Minimax value = +1

       Maximin value = -1

(ii)   $B_2$ dominates $B_3$ (Table S49)

### TABLE S49

| B<br>A | $B_1$ | $B_2$ |
|---|---|---|
| $A_1$ | -1 | 1 |
| $A_2$ | 3 | -4 |
| $A_3$ | -2 | 1 |

TABLE S50

| B:<br>A | B$_1$ | B$_2$ | Row<br>Min |
|---|---|---|---|
| A$_1$ | -1 | 1 | -1 ← |
| A$_2$ | 3 | -4 | -4 |
| Col<br>max | 3 | 1 | |

A$_1$ dominates A$_3$ giving Table S50 which cannot be reduced further.

(iii)    Minimax value = +1

Maximin value = -1

(iv)    There is no solution in pure strategies, since minimax $\neq$ maximin.

6.10(a)

TABLE S51

| B:<br>A | B$_1$ | B$_2$ | B$_3$ | Row<br>Min |
|---|---|---|---|---|
| A$_1$ | 1 | 2 | 3 | 1 |
| A$_2$ | 0 | 3 | -1 | -1 |
| A$_3$ | -1 | -2 | 4 | -2 |
| Col<br>max | 1 | 3 | 4 | |

There are no dominances, as can be seen from Table S51.    Maximin = minimax = 1 when A = A$_1$, B = B$_1$.    The value of the game is 1 to A.

(b)

TABLE S52

| B:<br>A | B$_1$ | B$_2$ |
|---|---|---|
| A$_1$ | a$_{11}$ | a$_{12}$ |
| A$_2$ | a$_{21}$ | a$_{22}$ |
| Col<br>max | a$_{11}$ | max (a$_{12}$, a$_{22}$) |

With reference to Table S52, suppose that a$_{11}$ is the saddle point. Then a$_{11}$ must be the row minimum of A$_1$, giving

$$a_{11} \leqslant a_{12} \tag{1}$$

Similarly a$_{11}$ must then be the column maximum of B$_1$, giving

$$a_{11} \geqslant a_{21} \tag{2}$$

Also, $a_{11} \leq \max \{a_{12}, a_{22}\}$, and $a_{11} \geq \min \{a_{21}, a_{22}\}$      (3)

We consider 2 cases:

*Case 1*, in which $a_{12} \geq a_{22}$

But $a_{11} \geq a_{21}$ (from (2)) , hence $A_1$ dominates $A_2$.

*Case 2* in which $a_{12} \leq a_{22}$

Then from (3) $a_{11} \leq a_{22}$; but $a_{11} \geq a_{21}$ (from (2) ), so $a_{21} \leq a_{22}$.
But from (1) $a_{11} \leq a_{12}$, hence $B_1$ dominates $B_2$.

(c)   The payoff matrix is as shown in Table S53.

### TABLE S53

| B: | 0 | 1 | 2 | 3 | Row Min |
|---|---|---|---|---|---|
| A: 0 | 0 | -1 | 0 | 1 | -1 |
| 1 | 1 | 0 | 0 | 1 | 0 |
| 2 | 0 | 0 | 0 | 1 | 0 |
| 3 | -1 | -1 | -1 | 0 | -1 |
| Col Max. | 1 | 0 | 0 | 1 | |

Since, for example if A bids £1 and B bids £0, A buys the ear-ring costing £2 for £1, making £1 gain. The minimax value of the game is 0 from B1 or B2 and the maximin value is 0 from A1 or A2. Hence the value of the game is zero, A and B bidding £1 or £2.

7.1     $V(B) = -V(A)$. $\beta$ is the value of the minimax to A, hence its value is $-\beta$ to B.

7.2 (a)   All the moves in $S_A^*$ are to be used, i.e. none of the relative frequencies can be zero.

(b)   Let $W_1$ be the value of the game to A if A adheres to $A_1$ completely, etc. Then $-V = P_1 (-W_1) + P_4 (-W_4) + P_7 (-W_7)$ = value of game to B.
Now, $W_1 \leq V, W_4 \leq V, W_7 \leq V$. Suppose $W_1 < V$, then $V = P_1 W_1 + P_4 W_4 + P_7 W_7 < (P_1 + P_4 + P_7) V = V$.
Contradiction, so $W_1 = V$. Similarly $W_4 = W_7 = V$.

7.3 (a)   It follows from result 1, i.e. $\alpha \leq V \leq \beta$.

(b)   This game has no saddle point, so $A_1, A_2, B_1, B_2$ must all be used, and $\alpha = -1$, $\beta = 1$, giving $-1 \leq V \leq 1$.

If V is value of game and $S_A^* = \left\{ \begin{array}{cc} A_1 & A_2 \\ P_1 & P_2 \end{array} \right\}$, $S_B^* = \left\{ \begin{array}{cc} B_1 & B_2 \\ q_1 & q_2 \end{array} \right\}$,

then $V = P_1 (-1) + P_2 (3)$   (B uses $B_1$)

and $V = P_1 (1) + P_2 (-4)$   (B uses $B_2$),

giving $P_1 = 7/9$, $P_2 = 2/9$, $V = -1/9$.

Also $V = q_1 (-1) + q_2 (1)$     (A uses $A_1$)

and   $V = q_1 (3) + q_2 (-4)$     (A uses $A_2$),

giving $q_1 = 5/9$, $q_2 = 4/9$, $V = -1/9$ (check).

(c)       $V = p_1 a_{11} + p_2 a_{21}$     (B uses $B_1$)

and   $V = p_1 a_{12} + p_2 a_{22}$     (B uses $B_2$),

giving $p_1 = (a_{22} - a_{21}) / (a_{11} + a_{22} - a_{12} - a_{21})$

and      $p_2 = (a_{11} - a_{12}) / (a_{11} + a_{22} - a_{12} - a_{21})$,

since $p_1 + p_2 = 1$.    Substituting, we obtain

$$V = (a_{11} a_{22} - a_{12} a_{21}) / (a_{11} + a_{22} - a_{12} - a_{21}).$$

Also $V = q_1 a_{11} + q_2 a_{12}$     (A uses $A_1$)

and   $V = q_1 a_{21} + q_2 a_{22}$     (A uses $A_2$),

giving $q_1 = (a_{22} - a_{12}) / (a_{11} + a_{22} - a_{12} - a_{21})$,

$q_2 = (a_{11} - a_{21}) / (a_{11} + a_{22} - a_{12} - a_{21})$

and the same value of V as above.

7.4 (a)   $A_2 V A_1$, since this diagram gives B's optimum strategies, the coefficients of B's gain matrix being the negatives of those for A.

(b)   $a_{12} = a_{21}$

(c)   See Figs S22 and S23.

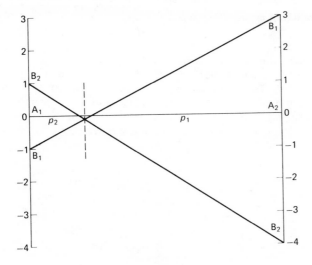

Fig. S.22

$$p_1 = .78 \simeq 7/9$$
$$p_2 = .22 \simeq 2/9$$
$$V = -.1 \simeq -1/9$$

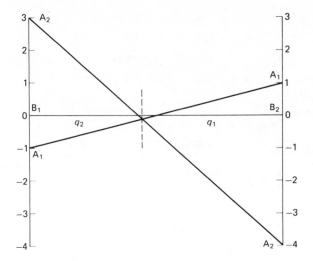

Fig. S.23

$$q_1 = .56 \simeq 5/9$$
$$q_2 = .44 \simeq 4/9$$
$$V = -.1 \simeq -1/9$$

7.5(a) If in table 32 on page 42 of the text we add another column $B_3$ with entries $a_{13}$, $a_{23}$, then we must add a further equation to the first set, corresponding to B playing $B_3$, i.e.

$$V = a_{13} \, p_1 + a_{23} \, p_2$$

So together with the other 2 equations we have 3 equations for the two variables $p_1$, $p_2$. These equations will have no solution unless there are only two independent equations (the solution to the first two must satisfy the third). Hence, in general, B will play just two strategies. In graphical terms B will play just two strategies unless 3 or more of the lines $B_i \, B_i$ pass through a single point.

(b) See Fig. S24.

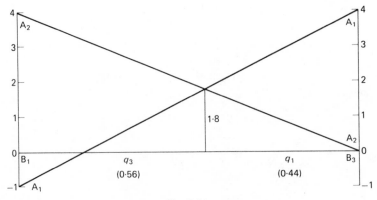

Fig. S.24

7.6(a)   $A_4$ dominates $A_1$ and $B_3$ dominates $B_2$, hence the game reduces to that of Table S54,

### TABLE S54

|       | $B_1$ | $B_3$ | Row Min |
|-------|-------|-------|---------|
| $A_2$ | 1     | 2     | ①       |
| $A_3$ | -1    | 3     | -1      |
| $A_4$ | 4     | 0     | 0       |
| Col max | 4   | ③     |         |

which has no saddle point.

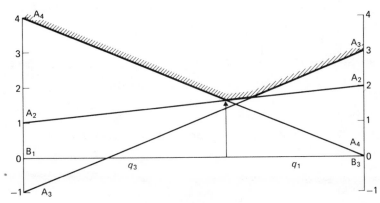

Fig. S.25

In this case B can change his proportions and A his strategies.  If B plays $B_1$ only (Fig. S25) then A will obviously play $A_4$;  if B takes $q_1$ to be 0.3, then A will play $A_3$ and the value of the game will be about 1.8.  Hence by suitable choices of strategy A can keep the value of the game to values along the shaded boundary.  B will try to minimise the value of the game, i.e. choose $q_1 \simeq 0.4$, $q_2 \simeq 0.6$ and the value of the game is $\simeq 1.6$ to A with A using $A_2$, $A_4$.  If we now plot the reduced game (Fig. S26, omitting $A_3$), we find $p_2 = 0.8$ and $p_4 = 0.2$.

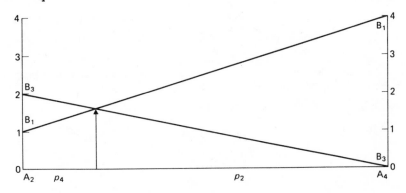

Fig. S.26

Hence the solution is

$p_1 = 0$, $p_2 = 0.8$, $p_3 = 0$, $p_4 = 0.2$,

$q_1 = 0.4$, $q_2 = 0$, $q_3 = 0.6$,

$V(A) = 1.6$

(b) From Table S55 we can see that for B strategy 5 dominates 1,2, and 4 since $b > a > 0$. Hence the game reduces to that shown in Table S56, which has no saddle point.

TABLE S55

| | | B | | | | |
|---|---|---|---|---|---|---|
| | | 1 | 2 | 3 | 4 | 5 |
| A | 1 | a | O | -b+a | O | O |
| | 2 | O | O | a | O | -b+a |

TABLE S56

| | 3 | 5 |
|---|---|---|
| 1 | -b+a | O |
| 2 | a | -b+a |

The solution in mixed strategies is

$$A\left\{\begin{array}{cc} 1 & 2 \\ b/(2b-a) & (b-a)/(2b-a) \end{array}\right\}, \quad B\left\{\begin{array}{ccccc} 1 & 2 & 3 & 4 & 5 \\ O & O & (b-a)/(2b-a) & O & b/(2b-a) \end{array}\right\}$$

with the value of the game being $(a-b)^2/(a-2b)$.

8.1 As suggested, put $V = -U$, $U > O$.

Set $x_1 = p_1/U$, $x_2 = p_2/U$, etc., so that we shall have $x_1 \geqslant 0$, $x_2 \geqslant 0$, etc.

Equation (1a) becomes $x_1 + x_2 + \ldots + x_m = U$

and (2a) $x_1 a_{1j} + x_2 a_{2j} + \ldots + x_m a_{mj} \geqslant -1$ for $j=1 \ldots m$

or equivalently $-x_1 a_{1j} - x_2 a_{2j} - \ldots - x_m a_{mj} \leqslant 1$ for $j=1 \ldots m$

Now, A wishes to maximise V,

or minimise U,

or maximise 1/U.

The problem may therefore be written as

Maximise $1/U = x_1 + x_2 + \ldots + x_m$

subject to $-x_1 a_{11} - x_2 a_{21} \ldots -x_m a_{m1} \leqslant 1$

.

.

.

$-x_1 a_{1n} - x_2 a_{2n} \ldots -x_m a_{mn} \leqslant 1$

and $x_1, x_2, \ldots, x_m \geqslant O$

Alternatively, we can write the result in terms of the elements of the matrix from B's point of view, since $b_{ij} = -a_{ij}$:

Maximise $1/U = x_1 + x_2 + \quad + x_m$

subject to $x_1 b_{11} + x_2 b_{21} + \ldots + x_m b_{m1} \leqslant 1$

.

.

.

$x_1 b_{1n} + x_2 b_{2n} + \quad + x_m b_{mn} \leqslant 1$

$$\text{and } x_1, \ldots\ldots\ldots, x_m \geqslant 0$$

8.2    Arguing as in Solution 8.1, we obtain

Minimise    $1/U = y_1 + y_2 + \ldots + y_n$

subject to    $y_1 b_{11} + y_2 b_{12} + \ldots + y_n b_{1n} \geqslant 1$

.
.
.

$$y_1 b_{m1} + y_2 b_{m2} + \ldots + y_n b_{mn} \geqslant 1$$

$$y_1 \ldots y_n \geqslant 0$$

8.3    We have two ways in which to complete this problem.

(i) The easier is to realise that the game has been reduced to a 2 x 3 game since A is not using $A_3$.

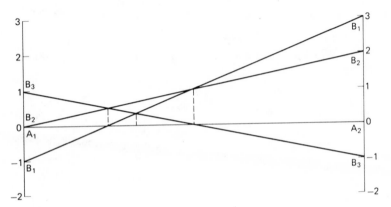

Fig. S.27

B will use $B_1$ and $B_3$ since he is trying to minimise $V(A)$ (Fig. S27), giving

$P_1 \quad = 2/3$

$P_2 \quad = 1/3$

$V(A) = 1/3$, as in the simplex solution, and from Fig. S28

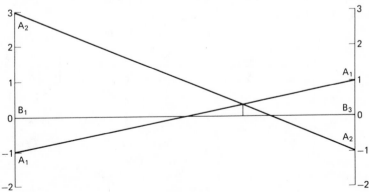

Fig. S.28

$V(A) = 1/3$

$q_1 \approx 1/3$

$q_2 \approx 2/3$

$q_3 = 0$

(ii) This problem can be completed using the dual L.P. problem, viz.

Maximise $1/V = y_1 + y_2 + y_3$

subject to

$$3y_1 + 4y_2 + 5y_3 \leqslant 1$$
$$7y_1 + 6y_2 + 3y_3 \leqslant 1$$
$$y_1 + 5y_2 + 4y_3 \leqslant 1$$
$$y_1, y_2, y_3 \geqslant 0$$

We obtain the tableau of Table S57 leading to the solution

$y_1 = 1/13$, $y_2 = 0$, $y_3 = 2/13$, $1/V = 1/13 + 2/13 = 3/13$

$(V = 13/3)$, $q_1 = 1/3$, $q_2 = 0$, $q_3 = 2/3$

This is the same result as obtained graphically, after subtracting 4 from V.

### TABLE S57

|   | $y_1$ | $y_2$ | $y_3$ | r | s | t |   |
|---|---|---|---|---|---|---|---|
| P | -1 | -1 | -1 | 0 | 0 | 0 | 0 |
| r | 3 | 4 | 5 | 1 | 0 | 0 | 1 |
| s | 7 | 6 | 3 | 0 | 1 | 0 | 1 |
| t | 1 | 5 | 4 | 0 | 0 | 1 | 1 |

8.4(a) (i) The inequalities are correct for a maximisation problem, hence the dual is

Minimise $Q = 10u + 20v + 16w$

subject to

$$u + v + 2w \geqslant 1$$
$$2u - v + w \geqslant 1$$
$$v + 4w \geqslant 2$$
$$u, v, w \geqslant 0$$

The original problem is probably the easier to solve.    The final tableau is as in Table S58.

### TABLE S58

|   | x | y | z | r | s | t |   |
|---|---|---|---|---|---|---|---|
| P | $\frac{1}{4}$ | 0 | 0 | $\frac{1}{4}$ | 0 | $\frac{1}{2}$ | $10\frac{1}{2}$ |
| y | $\frac{1}{2}$ | 1 | 0 | $\frac{1}{2}$ | 0 | 0 | 5 |
| s | $\frac{9}{8}$ | 0 | 0 | $\frac{5}{8}$ | 1 | $-\frac{1}{4}$ | $22\frac{1}{4}$ |
| z | $\frac{3}{8}$ | 0 | 1 | $-\frac{1}{8}$ | 0 | $\frac{1}{4}$ | $2\frac{3}{4}$ |

It follows that $u = \frac{1}{4}$, $v = 0$, $w = \frac{1}{2}$, $Q = 10\frac{1}{2}$ is the solution of the dual.

(ii) The inequalities have to be rearranged for the 'standard' minimisation problem

Minimise $\quad P = x_1$

subject to $\quad x_1 + x_2 + x_3 \geqslant 6$

$\qquad\qquad -x_1 - x_2 - x_3 \geqslant -6$

$\qquad\qquad x_1 \qquad\quad + 2x_3 \geqslant 4$

$\qquad\qquad x_1 - x_2 \qquad\qquad \geqslant 0$

$\qquad\qquad x_1, x_2, x_3 \geqslant 0$

The dual is then

Maximise $\quad Q = 6y_1 - 6y_2 + 4y_3$

subject to $\quad y_1 - y_2 + y_3 + y_4 \leqslant 1$

$\qquad\qquad y_1 - y_2 \qquad\quad - y_4 \leqslant 0$

$\qquad\qquad y_1 - y_2 + 2y_3 \qquad \leqslant 0$

$\qquad\qquad y_1, y_2, y_3, y_4 \geqslant 0$

The dual is probably the easier to solve: the final tableau is shown in Table S58, showing a degenerate solution with $y_1 = y_2 = y_3 = y_4 = 0$ and $Q = 0$. The solution of the dual is also seen to be degenerate with $x_1 = x_2 = 0$, $x_3 = 6$ and $P = 0$.

TABLE S59

|   | $y_1$ | $y_2$ | $y_3$ | $y_4$ | r | s | t |   |
|---|---|---|---|---|---|---|---|---|
| $Q$ | 0 | 0 | 8 | 0 | 0 | 0 | 6 |   |
| r | 0 | 0 | -3 | 0 | 1 | 1 | -2 | 1 |
| $y_1$ | 1 | -1 | 2 | 0 | 0 | 0 | 1 | 0 |
| $y_4$ | 0 | 0 | 2 | 1 | 0 | -1 | 1 | 0 |

(b) The original problem is

Minimise $\quad P - x_1 + x_2 \qquad\qquad\qquad - M(u + v) = 0$

subject to $\qquad\quad x_1 + x_2 - r \qquad\qquad + u \qquad = 5$

$\qquad\qquad\quad 2x_1 - x_2 \qquad - s \qquad\qquad + v = 4$

$\qquad\qquad\quad 3x_1 + 2x_2 \qquad\qquad + t \qquad\qquad = 6$

After putting the equations into standard form with respect to $t = u = v = 0$, we obtain the solution shown in Table S60.

The last tableau shows that the problem has no solution since the dummy variables cannot be eliminated.

The dual problem is

Maximise $\quad P = 5y_1 + 4y_2 - 6y_3$

subject to $\qquad y_1 + 2y_2 - 3y_3 \leqslant 1$

$\qquad\qquad y_1 - y_2 - 2y_3 \leqslant -1$

$\qquad\qquad y_1, y_2, y_3 \geqslant 0$

The second constraint has to be written as $-y_1 + y_2 + 2y_3 \geqslant 1$ before

*TABLE* S60

| | $x_1$ | $x_2$ | r | s | t | u | v | | |
|---|---|---|---|---|---|---|---|---|---|
| P | ⟨1−3M⟩ | −1 | M | M | O | O | O | −9M | |
| u | 1 | 1 | −1 | O | O | 1 | O | 5 | 5 |
| v | ⟨2⟩ | −1 | O | −1 | O | O | 1 | 4 | ②  |
| t | 3 | 2 | O | O | 1 | O | O | 6 | 2 |
| P | O | ⟨−½−3M/2⟩ | M | ½−M/2 | O | O | | −2−3M | |
| u | O | 3/2 | −1 | ½ | O | 1 | | 3 | 2 |
| $x_1$ | 1 | −½ | O | −½ | O | O | | 2 | NEG |
| t | O | ⟨7/2⟩ | O | 3/2 | 1 | O | | O | ⟨+O⟩ |
| P | O | O | MM | (5+M)/7 | (1+3M)/7 | O | | −2−3M | |
| u | O | O | −1 | −1/7 | −3/7 | 1 | | 3 | |
| $x_1$ | 1 | O | O | −2/7 | 1/7 | O | | 2 | |
| $x_2$ | O | 1 | O | 3/7 | 2/7 | O | | O | |

its insertion into the simplex tableau. The solution is as shown
(Table S61, after the insertion of a dummy). We see that the
problem has an *infinite* solution. This result is quite general:
the dual of a problem with an infinite solution is a problem with no
solution and vice versa.

*TABLE* S61

| | $y_1$ | $y_2$ | $y_3$ | r | s | t | | |
|---|---|---|---|---|---|---|---|---|
| P | −5+M | −4−M | ⟨6−2M⟩ | O | M | O | −M | |
| r | 1 | 2 | −3 | 1 | O | O | 1 | NEG |
| t | −1 | 1 | ⟨2⟩ | O | −1 | 1 | 1 | ⟨½⟩ |
| P | −2 | ⟨−7⟩ | O | O | 3 | | −3 | |
| r | −½ | ⟨7/2⟩ | O | 1 | −3/2 | | 5/2 | ⟨5/7⟩ |
| $y_3$ | −½ | ½ | 1 | O | −½ | | ½ | 1 |
| P | ⟨−3⟩ | O | O | 2 | O | | 2 | |
| $y_2$ | −1/7 | 1 | O | 2/7 | −3/7 | | 5/7 | NEG |
| $y_3$ | −3/7 | O | 1 | −1/7 | −2/7 | | 1/7 | NEG |

8.5 (a)

*TABLE* S62

| | $B_1$ | $B_2$ | $B_3$ |
|---|---|---|---|
| $A_1$ | 3 | 4 | −7 |
| $A_2$ | −2 | 1 | 6 |
| $A_3$ | 5 | 5 | −8 |

*TABLE* S63

| | $B_1$ | $B_3$ |
|---|---|---|
| $A_1$ | 3 | −7 |
| $A_2$ | −2 | 6 |
| $A_3$ | 5 | −8 |

We see from Table S62 that $B_1$ dominates $B_2$. Hence the problem
reduces to that of Table S63, which can be solved graphically
(Fig. S29) to find which of A's strategies are optimum.

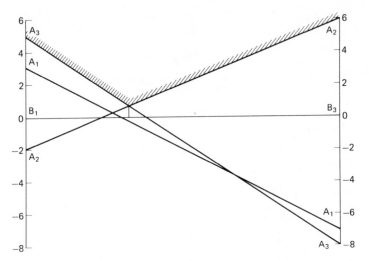

Fig. S.29

A will use $A_2$ and $A_3$ only, the value of the game being 2/3 to A. B uses $B_1$ with relative frequency 2/3 and $B_3$ the rest of the time. Now either by algebraic or graphical methods we find that A uses $A_2$ with relative frequency 13/21 and $A_3$ the rest of the time.

(b)    (i)   See Table S64.

### TABLE S64

B goes to

|  |  | X | Y | Z |
|---|---|---|---|---|
|  | X | 10 | -5 | -10 |
| A goes to | Y | -5 | 10 | -5 |
|  | Z | -10 | -5 | 10 |

(ii) There are no dominances and there is no saddle point.    Hence we must resort to linear programming.    To do this we must make V > 0.    This is most easily achieved by adding 15 (say) to all elements (Table S65).    Since each element is a multiple of 5, we can divide by this factor to simplify the arithmetic (Table S66).    It is easier to solve the problem from B's point of view by linear programming.    The *final* tableau is shown as Table S67.

### TABLE S65

| B: | X | Y | Z |
|---|---|---|---|
| A |  |  |  |
| X | 25 | 10 | 5 |
| Y | 10 | 25 | 10 |
| Z | 5 | 10 | 25 |

### TABLE S66

| B: | X | Y | Z |  |
|---|---|---|---|---|
| A |  |  |  |  |
| X | 5 | 2 | 1 | $p_1$ |
| Y | 2 | 5 | 2 | $p_2$ |
| Z | 1 | 2 | 5 | $p_3$ |
|  | $q_1$ | $q_2$ | $q_3$ |  |

## TABLE S67

|  | $y_1$ | $y_2$ | $y_3$ | $r$ | $s$ | $t$ |  |
|---|---|---|---|---|---|---|---|
| P | 0 | 0 | 0 | $3/22$ | $1/11$ | $3/22$ | $4/11$ |
| $y_1$ | 1 | 0 | 0 | $21/88$ | $-1/11$ | $-1/88$ | $3/22$ |
| $y_2$ | 0 | 1 | 0 | $-1/11$ | $3/11$ | $-1/11$ | $1/11$ |
| $y_3$ | 0 | 0 | 1 | $-1/88$ | $-1/11$ | $21/88$ | $3/22$ |

This leads to the solution

$$y_1 = 3/22 = q_1/(4/11) \qquad \text{giving} \qquad q_1 = 3/8$$
$$y_2 = 1/11 = q_2/(4/11) \qquad\qquad\qquad q_2 = 1/4$$
$$y_3 = 3/22 = q_3/(4/11) \qquad\qquad\qquad q_3 = 3/8$$

and the *true* value of the game $(11/4)\ 5 - 15 = -5/4$.

By using duality we see that

$$x_1 = 3/22 = p_1/(4/11) \qquad \text{giving} \qquad p_1 = 3/8$$
$$x_2 = 1/11 = p_2/(4/11) \qquad\qquad\qquad p_2 = 1/4$$
$$x_3 = 3/22 = p_3/(4/11) \qquad\qquad\qquad p_3 = 3/8$$

Hence the strategies for the two players are identical.

(c)  This problem is much more difficult to formulate than those given previously.  We have to consider all the possible ways in which Red can reply and eliminate those which are meaningless or stupid.  For example Red will never guess the same number twice!  Also if Blue chooses 3 and Red guesses 2, Blue's reply of 'low' will ensure that Red guesses correctly next time.  Bearing this in mind the payoff matrix is as shown in Table S68.  This matrix has neither dominance nor saddle, hence we must solve it by linear programming methods. One solution is as follows:

$$A \left\{ \begin{matrix} \text{i} & \text{ii} & \text{iii} & \text{iv} & \text{v} \\ 0 & 1/5 & 3/5 & 1/5 & 0 \end{matrix} \right\}, \qquad B \left\{ \begin{matrix} 1 & 2 & 3 \\ 2/5 & 2/5 & 1/5 \end{matrix} \right\}$$

and the value of the game (to A) is $1/5$.

## TABLE S68

|  |  |  | B | | |
|---|---|---|---|---|---|
|  |  |  | 1 | 2 | 3 |
|  | i | (1,2,3) | 1 | 0 | -1 |
|  | ii | (1,3,2) | 1 | -1 | 0 |
| A | iii | (2) | 0 | 1 | 0 |
|  | iv | (3,1,2) | 0 | -1 | 1 |
|  | v | (3,2,1) | -1 | 0 | 1 |

# Index